Astronomers' Universe

More information about this series at http://www.springer.com/series/6960

Gareth Wynn-Williams

Surveying the Skies

How Astronomers Map the Universe

 Springer

Gareth Wynn-Williams
Kailua, HI, USA

ISSN 1614-659X   ISSN 2197-6651   (electronic)
Astronomers' Universe
ISBN 978-3-319-28508-5   ISBN 978-3-319-28510-8   (eBook)
DOI 10.1007/978-3-319-28510-8

Library of Congress Control Number: 2016931900

Cover design: Courtesy of the European Space Agency

Printed on acid-free paper

This Springer imprint is published by Springer Nature
The registered company is Springer International Publishing AG Switzerland

Dedicated to Chris, Oliver, and Harry

Preface

I was never an amateur astronomer; foggy London in the 1950s was no place for a young boy to learn the constellations. But I did grow up as a science geek, surrounded by chemistry sets, construction kits, electric trains, and soldering irons, all strongly encouraged by my physicist father and math-teacher mother.

At Cambridge University in the 1960s I majored in physics and then looked around for a topic to pursue for my PhD. The decision was easy. Martin Ryle and his colleagues had built a major radio observatory at Lord's Bridge, a few miles outside of Cambridge, and data were pouring into its telescopes, rain or shine. With them we could see phenomena that were invisible to the largest optical telescopes in the world, and new discoveries were being made every month.

Having completed by PhD dissertation (on compact HII regions) I switched wavelengths and continents to join Gerry Neugebauer's group at Caltech, which was opening up the skies at infrared wavelengths. Once again, new science was pouring in through a newly opened window on the universe. I continued my research into infrared astronomy after I moved to the University of Hawaii in 1978.

One of the most important things I learned from all this is the way that new technology is a driving force of astronomy. When a new technology is invented, progress tends to come in three stages: first, some exciting discoveries are made by looking at a few favorite places in the sky. Then someone does an initial survey of the sky and starts to look for patterns and statistics in the new data. Finally, major surveys are undertaken to build databases that astronomers everywhere can access and search for new phenomena. This is the idea that has guided the structure of this book.

It is a pleasure to thank the following people who helped me find images, or gave me feedback on various parts of the book: Charles Alcock, Eric Becklin, Bruce Berriman, Alec Boksenberg, Thomas Dame, Gary Davis, Larry Denneau, David DeVorkin, Alison Doane, Graham Dolan, Yasuo Doi, Steve Drake, Tom Donaldson, Michael Endl, Christopher Erdmann, Bill Forman, Gordon Garmire, Michael Geffert, Jonathan Gradie, David Green, Adam Hart-Davis, Pat Henry, Stefan Hughes, Bob Joseph, Norbert Junkes, Bill Kendrick, Tom Kerr, Robert Kirshner, Brandon Lawton, Anthony Lasenby, Charles Lawrence, Malcolm Longair,

Gene Magnier, Axel Mellinger, Chris Oliver, Alex Ostrovsky, Mike Perryman, Guy Pooley, Tony Readhead, Dave Sanders, Fred Seward, Wendy Stenzel, Woody Sullivan, Gary Thompson, Alan Tokunaga, John Tonry, Scott Wakely, Gavin White and Glenn White.

I started writing this book soon after my retirement from the University of Hawaii. I have benefitted enormously from subsequently being provided with a desk, library access, and intellectual stimulation both at the Institute for Astronomy in Honolulu and at the Cavendish Astrophysics group at Cambridge University. Particular thanks go to directors Guenther Hasinger and Paul Alexander for these privileges.

A toast is also due to Wikipedia founder Jimmy Wales and the dozens of anonymous contributors who have made Wikipedia such a superb first resource for astronomy. I have no shame in revealing that Wikipedia is usually the first place I visit when I need to explore an astronomy topic that is new to me. Thanks, too, to the many astronomers and science writers, particularly at NASA, who maintain the splendid public information websites associated with many of the projects mentioned in this book.

Honolulu, HI, USA and Cambridge, UK   Gareth Wynn-Williams
April 2016

Contents

Chapter 1
Introduction

There is no way to know who first surveyed the sky: it may not even have been a human. From experiments in planetaria we know that migrating birds recognize the rotation of the sky, and it has been reported that dung beetles can navigate by the light of the Milky Way.

With the naked eye we can see the Sun, the Moon, five planets, and around 1,000 stars. Although the nighttime sky's appearance changes from hour to hour and month to month, the relative positions of the bright stars hardly change over generations. The constellation patterns we see today are almost identical to those observed by the ancient Babylonians thousands of years ago, and there are stars that have had the same name for millennia.

But while the sky doesn't change much, the tools that are available to astronomers do. Ever since Galileo first used his telescope, astronomers have been searching out new ways to scan the heavens, and each new technology has led to the discovery of new kinds of object. The telescope led to the discovery of nebulae, radio telescopes led to the discovery of quasars, infrared astronomy gave us protostars, while X-ray astronomy revealed black holes.

Most of present-day professional astronomy is actually astrophysics. We focus on finding out what things are made of, how they became the way they are, how they interact with each other, and how they will behave in the future. But pure astronomy still exists in the form of the search for new objects and new phenomena, and it is by conducting new kinds of surveys of the sky that our science is continually being reborn.

© Springer International Publishing Switzerland 2016
G. Wynn-Williams, *Surveying the Skies*, Astronomers' Universe,
DOI 10.1007/978-3-319-28510-8_1

1.1 The Five Eras of Astronomy

There are many different ways to describe the history of astronomy, but a particularly useful one is to focus on new technology as a major trigger for progress. This approach leads us naturally to divide astronomy into five distinct historical eras. When a new era begins there is a need and an enthusiasm to survey the sky by some new method.

The Naked Eye Era This goes from prehistoric times up to the invention of the telescope in the 17th century. This era is covered in Chapter 2.

The Telescope Era During this era, which began with Galileo, astronomers could look at the sky through telescopes, but had to record what they saw by hand. The number of catalogued stars rose from about a thousand to nearly a million. Asteroids, nebulae, and planetary moons were added to the pantheon of cosmic phenomena. This era is covered in Chapter 3.

The Photography Era The third era is dominated by the introduction of astrophotography around 1880, allowing astronomers to obtain permanent records of what they saw, and to detect much fainter objects than those visible through an eyepiece. The greatest events in the photography era were the discovery of the nature of galaxies, and the understanding of the physics of stars. The photography era is covered in Chapter 4.

The Electromagnetic Era Until Jansky's radio experiments in the 1930s all astronomy was done using visible light, except for small extensions into the ultraviolet, and infrared regions that were made possible by special photographic plates. Radio astronomy blossomed soon after World War II, but the opening up of the X-ray, gamma-ray, ultraviolet and infrared wavebands started in the 1960s, largely as a by-product of the space race. Once it was established that there was science to be done in these wavebands, astronomers built instruments to survey the sky, pinpointing objects that were worth studying in more detail with specialized telescopes. These surveys are covered in Chapters 5–10. The space age also facilitated two vastly improved astrometry surveys, which are described in Chapter 11.

The CCD Era With the ever-increasing sophistication of solid-state digital cameras, and the ever-decreasing cost of digital technology, we are now firmly in the fifth era of astronomy, where sources are being discovered by the billions mainly, but not exclusively, by once again using ground-based telescopes at visible wavelengths. Telescopes fitted with large digital cameras, such as Pan-STARRS in Hawaii, are programmed to methodically and repeatedly scan the sky and record their results in vast digital databases. Part of the motivation for these surveys is to look for moving and changing objects in the sky, such as potentially hazardous asteroids, but the databases are also gold mines for the discovery and study of stars and galaxies. More and more astronomy is now being performed by making digital searches within these databases rather than traveling to a mountain-top observatory to collect new data. This era is covered in Chapter 12.

1.2 Defining a Survey

There are five main factors that define the properties of an astronomical survey: wavelength, sky coverage, angular resolution, sensitivity, and time-variability. Each of the surveys described in this book has involved a major innovation or improvement in at least one of these factors.

Wavelength The single most important factor in a survey is the wavelength at which it is conducted. As Figure 1.1 shows, there is now an enormous range of electromagnetic waves that are accessible to astronomers. The ratio between the longest and shortest wavelength observations is now a factor of 10^{21}. In the photographic era the ratio was about 3, and in the naked-eye era it was less than 2.

Since the Earth's atmosphere is opaque over wide ranges of the electromagnetic spectrum the choice of wavelength will usually determine whether the survey will be conducted from the ground or from a space-borne telescope. All X-ray observations, plus many infrared, ultraviolet, and gamma-ray observations, must be made from space, but most visible-light and radio observations can be done from the ground at a much lower cost.

A survey will often be performed simultaneously at two or more wavelengths, because the ratio of an object's brightness at different wavelengths, which we would call the color in the case of visible light, is often a powerful guide to the physical nature of the object under study.

Sky Coverage On a cloudless night at sea-level a telescope can view half of the universe: the other half is below its horizon. As the Earth rotates and orbits the Sun other parts of the sky become temporarily visible, but unless the telescope is situated exactly on the equator there will always be parts of the universe that never rise above its horizon. Many surveys accept this limitation and confine themselves to more or less half of the sky, but there are a few, including the 2MASS survey (section 6.2), which employ two telescopes—one in each hemisphere.

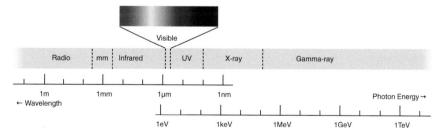

Fig. 1.1 The electromagnetic spectrum currently available to astronomers. As discussed in Appendix A.3, astronomers usually use photon energy rather than wavelength to describe X-rays and gamma rays

Another solution to the horizon problem is to go into space. An Earth-orbiting survey telescope that spends part of its orbit in the northern hemisphere and part in the south can potentially observe the whole sky, so long as it stays in orbit long enough that the Sun is not always in front of the same constellation. A few satellites, including WMAP (section 7.3), have been placed into a special Lagrangian orbit that takes them around the Sun rather than the Earth.

Other factors may come into play when considering the sky coverage of a survey. If the survey's main purpose is to study distant galaxies, as is the case with the Sloan Digital Sky Survey (section 12.1), there is not much point in making observations in the direction of the Milky Way. Likewise, most searches for asteroids will be concentrated in the plane of the ecliptic.

Most of the surveys in this book cover at least a quarter of the sky, but in Chapter 13 we look at a few specialized surveys that made important discoveries based on studies of very small patches of sky,

Angular Resolution The term angular resolution refers to the sharpness of an astronomical image. Roughly speaking, it is the finest detail that can be discerned and is expressed as an angle on the sky. The angular resolution of a survey is sometimes referred to as the beamwidth; generally speaking, the smaller the beamwidth, the more information available from the observations. Under good conditions the naked eye can see details down to about one minute of arc—1/60 of a degree or 1/30 of the diameter of a full moon. The best ground-based telescopes can see details a bit smaller than one second of arc—1/3600 of a degree.

There are four major factors that can limit the angular resolution of a survey:

- **Diffraction** is a fundamental limitation based on the wavelength of the radiation being studied and on the diameter of the telescope being used: it is explained in more detail in Appendix A.4. The diffraction limit is a particularly serious problem for radioastronomy.
- **Atmospheric Seeing** is the result of starlight being deflected by random refractive index irregularities in the Earth's atmosphere. Its effect can be reduced by locating telescopes on good mountain-top sites, or eliminated by going into space. It is sometimes possible to improve the angular resolution of a ground-based telescope by a process called adaptive optics.
- **Telescope Optics:** In some cases surveys are limited by the precision of the telescope optics; this is the most common limitation of X-ray telescopes.
- **Detector Design:** In some surveys the resolution is limited by the sizes and numbers of detector used. The limitation might be due to the cost of the detectors or to the maximum data-rate that can be recorded on a strip chart or transmitted to the ground from a satellite.

Sensitivity The sensitivity of a survey sets a limit on the faintest objects than can be detected. The limiting sensitivity can be set by one of several factors:

- **Detector Noise:** There may be unavoidable electrical noise in the detector system. This noise can often be reduced by cooling the detectors, in some cases with liquid helium.
- **Background Noise:** Background noise can arise in numerous ways: the noise might be astronomical—from the interstellar medium or the cosmic background radiation—or it might be from the Earth's atmosphere, or it may be heat radiation from the telescope mirror itself.
- **Photon Noise:** The laws of physics state that electromagnetic energy cannot be measured in units smaller than a photon. As will be described in Chapter 10, this limitation is particularly serious in the case of gamma-ray astronomy.
- **Confusion:** A survey may detect so many objects that their images start to overlap. This problem is more severe if the angular resolution is poor. If the main purpose of a catalog is to throw up new objects for further study then a little bit of confusion does not greatly matter. But if the survey is to be used for statistical studies, astronomers will usually limit the catalog to objects above a certain brightness (the "confusion limit") so as to reduce the chances of images overlapping.

Variability Some astronomical surveys are specifically designed to look for sources that vary in brightness or some other property. The timescale for variation can be anything between milliseconds and decades. Examples include the Harvard Plate Collection (section 4.2), which focused on variable stars; the radio survey which discovered pulsars (section 5.5); the BEPPOSAX satellite which looked for gamma-ray bursts (section 10.4); and Pan-STARRS, which searches for moving objects such as asteroids (section 12.2).

Chapter 2
The Naked Eye Era

In the centuries before crosswords and computer games, trying to understand the motions of the stars and planets across the sky provided the best intellectual challenge that a curious mind could face.

Some changes, such as the day-night cycle, are easy to describe and predict, while others, such as the retrograde motions of the planets, are not. Subtleties such as the changing speed of the Moon's motion require careful observation, while the changing length of the seasons is obvious to anyone who tries to grow food.

There are many excellent books concerned with the early history of astronomy. The main focus of most of these books is our ancestors' attempts to understand the motions of the Sun, the Moon, and the five naked-eye planets against the much steadier background of stars. The works of Claudius Ptolemy and of Nicolaus Copernicus, two of the greatest astronomers ever, are almost entirely concerned with visualizing and predicting the paths of these nearby objects.

In this book we can skip rapidly over much of these parts of our history, since our focus is on the discovery of new objects rather than the behavior of those that have been known since prehistoric times.

But let us not forget our debt to the Moon and planets: the quest to understand their motions encouraged logical thought, pushed mathematics to new limits and inspired Isaac Newton to invent what we now call physics. If the Moon and planets of our solar system had not existed, or if they had been too faint to be seen with the naked eye, or if the earth's skies had always been cloudy, the development of all the sciences and much of civilization would surely have been delayed by many centuries.

© Springer International Publishing Switzerland 2016
G. Wynn-Williams, *Surveying the Skies*, Astronomers' Universe,
DOI 10.1007/978-3-319-28510-8_2

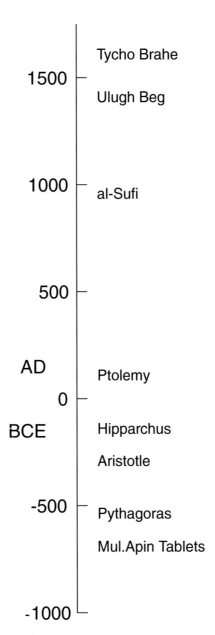

Fig. 2.1 Timeline for the naked eye era

Fig. 2.2 Mul.Apin clay tablet showing text on both sides

2.1 The Babylonians and the Mul.Apin Tablets

The oldest written star catalog we know of dates from the Babylonian era, about the 8th century BCE. It takes the form of small clay tablets crammed with cuneiform text (Figure 2.2). Several nearly identical groups of tablets have been found which describe the constellations that were identified by the Babylonians. They are referred to as the Mul.Apin tablets, after the first words of the texts.

Besides containing extensive descriptions of how the night sky changes over the months and seasons, the Mul.Apin tablets include a catalog of 71 constellations and stars, most of which have been identified with their modern counterparts (Figure 2.3). Some of these Babylonian constellations are direct precursors of those still in use today. For example we can easily recognize the descriptions of a scorpion (Scorpio), of twins (Gemini), and of a bull (Taurus). Most of the features listed in the Mul.Apin catalog are groups of stars, but there are a few individually identifiable stars, such as Arcturus and Sirius. The preference of the Babylonians for naming constellations rather than stars is perhaps a clue to understanding ancient attitudes to them. Nowadays we are used to thinking of a star as a hot ball of gas, perhaps with planets orbiting it. But to the Babylonians, and the Greeks that followed them, the stars had no more physical reality than the spots of light projected onto the dome of a planetarium. Their unchanging patterns—the constellations—were far more interesting to the ancients than were the stars themselves, both as markers for the motions of the planets and as inspirations for their cultural myths.

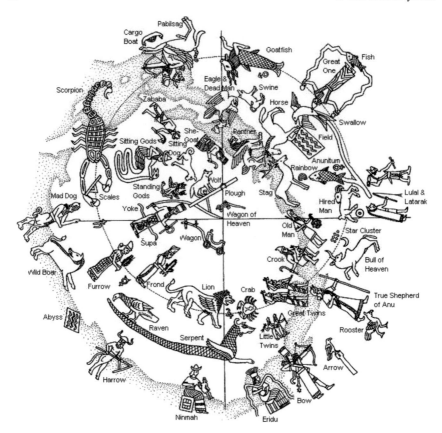

Fig. 2.3 Modern reconstruction of the Babylonian star map. Image credit: from 'Babylonian Star-Lore. An Illustrated Guide to the Star-lore and Constellations of Ancient Babylonia' by Gavin White, ©Solaria Publications 2014

2.2 The Greeks: Aristotle, Hipparchus, and Ptolemy

The Babylonians worked hard to come up with mathematical rules that would allow the seasons and Moon's motion to be fairly accurately predicted, but they do not seem to have felt the need to come up with a physical model for the universe.

The Greeks, on the other hand, tried to interpret what they saw, and formulated a three-dimensional picture of the universe. While the Babylonians and early Greeks visualized the Earth as flat, by the time of Pythagoras (570–495 BCE) several pieces of evidence, such as the shape of the Earth's shadow during a lunar eclipse, indicated that the Earth had to be a sphere. Aristotle (384–322 BCE) built on this idea and formulated a picture of the universe based on a stationary Earth around which the Sun, Moon, and planets orbited at different distances (Figure 2.4). Beyond the

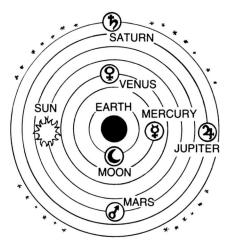

Fig. 2.4 The Aristotelian universe, with the Earth at its center and all the stars in a spherical shell around it

planets were the "immutable stars" in a spherical shell, all at the same distance from the Earth. This model, elaborated by Claudius Ptolemy several centuries later, survived for nearly two millennia until the time of Copernicus and the acceptance of the idea that the planets revolve around the Sun, not the Earth.

The earliest survey we know of that recorded actual numerical positions of stars was that of Timocharis (320–260 BCE) who listed the coordinates of 18 stars. We no longer have his data, but his measurements were used and discussed by several later astronomers. We know much more about the work of the great astronomer Hipparchus (190–120 BCE) who meticulously catalogued the positions of 850 stars, This number corresponds to most of the stars that can be seen with the naked eye from a European latitude.

Hipparchus developed and used some kind of armillary sphere (Figure 2.5) to make his measurements and achieved a positional accuracy of about half a degree— the diameter of the Earth's moon. The accuracy of Hipparchus' star positions allowed him to make several important discoveries. He came up with models for the motions of the Sun and the Moon and used these to predict eclipses. Most famously, by comparing his own measured positions for certain stars with those made by Timarchus 150 years earlier he discovered precession—the slow change in the direction of the Earth's rotation axis (see Appendix A.5). He also invented trigonometry, measured the length of the year to within seven minutes, and invented the magnitude system for describing the brightness of stars (see Appendix A.6).

Unfortunately, we do not possess any original copies of Hipparchus' star catalog, but there is good evidence that Ptolemy's star catalog in the Almagest, published some 250 years later, relies heavily on Hipparchus' data. The evidence for this assertion is based on the fact that the positions of the stars in Ptolemy's catalog differ from what one would expect if he had made his own observations in the second

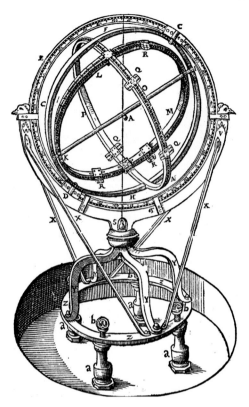

Fig. 2.5 This armillary sphere was built and used by Tycho Brahe, but is probably broadly similar to the one developed and used by Hipparchus. Measuring the position of a star with this device involves carefully aligning one of the brass rings along a North-South line through the zenith, and another to the ecliptic—the apparent path of the Sun through the sky. The star's position can then be read off the scales drawn on the appropriate rings

century AD. However, they are just what one would expect if Ptolemy had used Hipparchus's 250-year old data and precessed them using what we now know to be an incorrect formula. We can criticize this action of Ptolemy, and we can criticize his advocacy of the geocentric theory of the universe, but we should not lose sight of the fact that his major treatise, the Almagest, stands as one of the most important books ever written—the bedrock of astronomy and science for more than a thousand years.

Neither Hipparchus nor Ptolemy left us with a two-dimensional map of the sky, but Figure 2.6 shows a 2nd-century AD Roman copy of a Greek statue of the god Titan holding up the heavens on his shoulders. The surface of the globe contains illustrations of over 40 identifiable constellations as well as lines that correspond to the equator, the ecliptic, and the Arctic and Antarctic circles. This is the oldest known visual representation of the sky in existence, though there is controversy over whether the original dates from the time of Hipparchus or the time of Ptolemy.

Fig. 2.6 The statue of Titan holding a sky globe at the National Museum of Archaeology in Naples, Italy. It is sometimes referred to as the "Farnese Atlas." Image credit:Wikimedia

2.3 Islamic Astronomy

Hipparchus's sky survey, as incorporated into the work of Ptolemy, ruled unchallenged for a thousand years, but improvements came with the rise of Islamic astronomy in the 10th century AD.

The Book of Fixed Stars (see Figure 2.7), written by the Persian astronomer Abd al-Rahman al-Sufi around 964 AD, was primarily an attempt to introduce Ptolemaic ideas to the Muslim world, but it broke new ground in two important ways. First, it gave every star a name, rather than just a location within a constellation. Many of these Arabic names are still in everyday use today, including Aldebaran, Betelgeuse, Deneb, Fomalhaut, Rigel, and Vega. Second, it contains the earliest recorded observations of what we now know to be external galaxies, namely the Andromeda galaxy, to which he gave the description "little cloud," and the Large Magellanic Cloud. In making the latter discovery, al-Sufi had the advantage over Ptolemy of living at a lower geographic latitude, since the Magellanic Clouds are too far south to be visible from Greece.

Nearly 500 years later, Ulugh Beg, the governor of Samarkand, founded a Madrasah, or university, dedicated to the study of astronomy. He also set in motion the construction of what was probably the first observatory in the world to involve permanently constructed instruments, the largest of which was a giant 40-meter radius quadrant (see Figures 2.8 and 2.9). With this instrument Ulugh Beg and his colleagues produced a star catalog that had about the same number of stars as Ptolemy's but better accuracy and fewer errors. Those parts of the observatory which were built above ground were destroyed in 1449, but the section of the large sextant that had been built underground survived and was rediscovered in 1908. The observatory has since been reconstructed as a museum. Ulugh Beg was also a great mathematician, producing tables of sines and tangents that were accurate to eight decimal places, and measuring the length of a year to within half a minute.

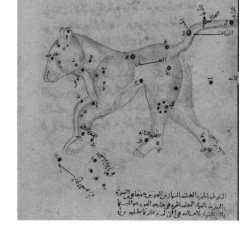

Fig. 2.7 The Ursa Major constellation as mapped by al-Sufi. This image is taken from the oldest extant copy of his Book of the Fixed Stars, dated around 1009 AD

Fig. 2.8 Reconstructed buildings of Ulugh Beg's observatory in Samarkand, now part of Uzbek-istan. This building houses the quadrant seen below. Light enters the quadrant though the doorway on the left of the picture. Image credit: Wikipedia

Fig. 2.9 The giant 40-meter radius quadrant that was used to measure the positions of stars and planets at Ulugh Beg observatory. Image credit: Alex Ostrovsky

2.4 Chinese Astronomy

Although ancient Chinese astronomers never attained the level of mathematical precision reached by Greek and Islamic astronomers, they paid great attention to mapping the positions of the stars in the sky. Figure 2.10 shows a small part of an extensive map of the sky discovered in 1907 in a cave in Dunhuang, on the Silk Road in western China. The map dates from somewhere between the 7th and 10th century AD, and is the oldest known paper-based map of the sky. It contains about 1,500 stars, considerably more than Ptolemy's Almagest.

Chinese astronomers also seem to have paid more attention to transitory phenomena than their western cousins, keeping extensive records of comets, novae, and even sunspots. In fact, the oldest astronomical records we know of are Chinese; they include oracle bones that refer to a solar eclipse that is known to have occurred in 1281 BCE, and to either a nova or a supernova that occurred about 1300 BCE.

Some of these records have turned out to be of great astrophysical significance, notably the report of a "guest star" in AD 1054, which many centuries later was identified as the supernova explosion that produced the Crab Nebula. Knowledge of the time since that explosion, plus the descriptions of how the brightness of the new star varied from day to day, have been of great value in understanding the physics of this remarkable object.

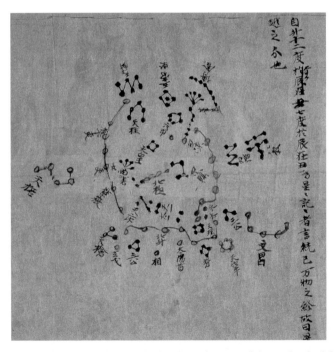

Fig. 2.10 Part of the Dunhuang star map showing the region around the north celestial pole

2.5 Tycho Brahe

Hipparchus's sky survey, as incorporated into the work of Ptolemy and Ulugh Beg, saw few major refinements until Tycho Brahe (1546–1601) decided to devote his career to astrometry—the precise measurement of star positions. Tycho was inspired by two celestial events early in his career; the appearance of bright new star in 1572, and of a comet in 1577.

Aristotle's view that the stars and the constellations were forever unchanging still held sway in Tycho's time; anything that moved or fluctuated in brightness was assumed to be occurring within the Earth's atmosphere. The new star of 1572, for which Tycho coined the word nova but which we would now classify as a supernova, gave him a chance to test this assumption. He realized that if the new star was closer than the Moon, as widely believed, he should see parallax as the Earth's daily rotation carried him from east to west. And if it belonged to the realm of the planets he should also see proper motions from day to day as the object moved in its orbit. The fact that he could detect neither kind of motion during the whole year that he could observe it indicated to him that the supernova belonged to the realm of the distant stars; it therefore violated Aristotle's premise that the stars were immutable. Several years later he applied the same test to the bright comet that lit up the skies in 1577; he showed that it was moving at a rate consistent with it being in a planet-like orbit—far above the Earth's atmosphere where comets were then supposed to exist—but much closer than the fixed stars.

Emboldened by this demonstration of the scientific value of precise astrometry, he sought and found a rich sponsor to finance his research activities. At Hven, in what was then Denmark but is now Sweden, he built the Uraniborg Observatory—a spacious building for himself and his staff that incorporated a number of specially designed angle-measuring instruments (Figure 2.11). All of the instruments required human observations with the naked eye: Tycho died several years before the telescope was invented.

One of the ways in which he and his assistants made their observations is shown in Figure 2.11. The large, precisely-engraved quadrant in the foreground was aligned exactly in a north-south direction and was used to measure the angle of a star above the horizon when it transited the meridian, as well as the precise time at which it did so. To achieve this, the bearded man at the far right of the picture faces the wall to the south of the quadrant which contains a small window with a pair of crossed wires in it, one vertical and one horizontal. As the star approaches the center of the window he adjusts the position of a sliding marker on the quadrant so that the star lines up with the horizontal wire in the window. He then watches the star move horizontally across the window and shouts when it crosses the vertical line. The man at the bottom right then reads out the time on the clock and the scribe at the lower left writes down both the time and the angle of the quadrant. From these two numbers the coordinates of the star can be calculated.

Fig. 2.11 Tycho Brahe at his Uraniborg observatory showing several of the instruments he used, including the large quadrant and the clock in the foreground

The first draft of Tycho's star catalog, circulated in 1598, contained 1004 stars with a positional accuracy of about half a minute of arc—sixty times better than Hipparchus. It was the first catalog to include corrections for atmospheric refraction—the bending of light as it passes though layers of different density.

At the same time as he was compiling his star catalog, Tycho and his staff carefully recorded the ever-changing positions of the Moon and planets against the background of the fixed stars. It is for these observations that we owe our greatest debt to him; his one-time assistant Johannes Kepler pored over these data for many years trying to find the patterns that governed the orbits of the planets. In 1609 he published his discovery that the planets move in elliptical orbits around the Sun contradicting once and for all the Ptolemaic insistence that all celestial motions were based on circles. Kepler did not try to provide any physical explanation for his laws of planetary motion, but eighty years later Isaac Newton proved that they are a mathematical consequence of his three laws of Motion plus his Universal Law of Gravity. If Tycho Brahe had not measured the positions of the planets as accurately as he did, Kepler would not have been able to show that Ptolemy's circle-based theory did not fit the observations satisfactorily. We must therefore thank Tycho Brahe and his meticulous astrometry for providing the secure foundation for what we now refer to as Classical Physics.

A few years later, in 1603, the German astronomer Johann Bayer used Tycho's catalog to create a star atlas called Uranometria (Figure 2.12). This atlas broke ground in two ways. First, it covered the whole sky, making use of the southern hemisphere observations of about 300 stars by the Dutch navigator Frederick de Houtman. Second, it introduced the constellation-based star naming system that is still in use today. In this system, stars within a constellation are named with Greek letters in order of decreasing brightness. Thus Betelgeuse, the brightest star in the Orion constellation is named α Orionis while the second brightest star, Rigel, is named β Orionis.

Fig. 2.12 The Orion region from Johann Beyer's Uranometria, showing the first use of Greek letters to describe the relative brightness of the stars in a constellation

Chapter 3
The Telescope Era

The second great era of astronomy runs from 1609, when Galileo first trained his telescope on the skies, to around 1880 when photography became a serious tool for astronomical research. As we shall see in this chapter, astronomers looking directly through the eyepieces of telescopes discovered new planets, planetary moons, asteroids, binary stars, variable stars, gaseous nebulae, star clusters, proper motions, and parallax. Perhaps most importantly, this was the era in which we recognized that the stars were objects of interest in their own right, and started measuring their spectra, their distances, and their motions with respect to each other.

The first telescope that we know about was built by the Dutchman Hans Lippershey in 1608, but we do not know if he ever trained his instrument on the sky. But by the following year Galilei Galileo of Pisa in Italy had built his own telescope and started to make written methodical records of what he saw. The science of astronomy was changed forever. Galileo discovered mountains on the Moon, resolved the Milky Way into individual stars, and correctly explained the nature of sunspots. Most importantly, he discovered moons around the planet Jupiter (see section 3.1) and was able to resolve the illuminated disk of the planet Venus, noticing how its varying size and shape could only be explained if its orbit took it behind the Sun. As Galileo realized, these two discoveries provided crucial support for Copernicus's still controversial heliocentric theory of the solar system, as opposed to Ptolemy's geocentric theory, a view that nearly cost him his life.

Improvements in telescope design over the next three centuries facilitated two quite different kinds of survey. On one side were the surveys that led to the discovery of new kinds of object—most notably nebulae and small bodies of the solar system. On the other side were the surveys whose aim was greater astrometric precision; as well as supporting celestial navigation, these high-precision surveys led to several major astrophysical discoveries, including parallax, proper motions, and binary star systems.

© Springer International Publishing Switzerland 2016
G. Wynn-Williams, *Surveying the Skies*, Astronomers' Universe,
DOI 10.1007/978-3-319-28510-8_3

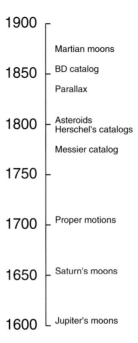

Fig. 3.1 Timeline for the telescope era

3.1 Surveying the Solar System

Prior to the invention of the telescope, the only known permanent members of the solar system were the Sun, the Earth, the Moon, and the five planets Mercury, Venus, Mars, Jupiter and Saturn. There were also the comets, which Tycho Brahe had established to be at planetary distances, but their nature and their orbits were not yet understood. Since then, successive generations of telescopes have led to the discovery of tens, then hundreds, and now millions of smaller members of our solar system. Some of these have been discovered as a result of careful surveys, while others have been found serendipitously.

Moons and Rings The first new solar system objects to be found were the four largest moons of Jupiter. Galileo announced their discovery in 1610, less than a year after his first attempts to construct a telescope. His methodical recording of what he saw each night (see Figure 3.2) convinced him that the four objects were orbiting Jupiter with periods of between 1.8 and 17 days. He referred to them as Mediciean Stars in honor of his patron; it was Kepler who later named them satellites—a word used to describe a subordinate servant of a powerful dignitary. They soon became known as Io, Europa, Ganymede, and Callisto. The discovery of objects that orbited a planet other than the Earth was an important piece of evidence against the geocentric theory of the universe that was the standard dogma at that time.

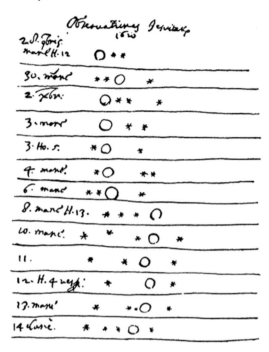

Fig. 3.2 Galileo's nightly recordings of Jupiter and its four surrounding moons. Compare these sketches with the recently-obtained image in Fig. 3.3

Fig. 3.3 Jupiter and its major moons as seen through a modern 25-cm telescope. Image credit: Jan Sandberg

Galileo's discovery spurred further searches, but the moons of other planets are considerably fainter than those of Jupiter and had to wait for improvements in telescope technology. Christiaan Huygens discovered Saturn's moon Titan in 1655 and Giovanni Cassini discovered four more moons of Saturn between 1671 and 1684. Mars's two very faint moons, Phobos and Deimos, were not detected until 1877. Even the most sensitive modern searches have failed to find any moons orbiting Mercury or Venus, though deep space missions in the 20th century have yielded many more small objects around Jupiter, Saturn, Uranus, Neptune, and even Pluto.

When he pointed his telescope at Saturn, Galileo noticed bright areas on each side of the planet, but was mystified as to their nature. Huygens, in 1655, with his much better telescope, recognized that what Galileo had seen was a flat ring surrounding Saturn. A few years later, Cassini resolved the disc into a number of separate rings. The detection of the very much fainter rings around Jupiter, Uranus, and Neptune had to wait until the space age.

Planets Only two new planets were discovered during the telescope era: Uranus and Neptune. Uranus was recognized as a planet by William Herschel in 1781, during his survey for nebulae (see section 3.3). He noted its extended appearance and its slow motion among the stars and initially believed that he had found a comet; its true nature as a new planet was clarified when the Russian astronomer Anders Lexell determined that its orbit around the Sun was nearly circular. A few years later, Herschel discovered the two largest moons of Uranus; Titania, and Oberon.

Neptune's discovery happened quite differently: observations of Uranus's motion round the Sun over the sixty years following its discovery indicated that it was being gravitationally tugged by an object that was probably a planet beyond Uranus. The French astronomer Urbain LeVerrier predicted the position of this new planet so well that it was discovered on the first night it was searched for in 1846. Its largest moon, Triton, was discovered less than three weeks later.

Asteroids Herschel's discovery of Uranus seemed to provide support for an empirical relationship called the Titus-Bode Law. The law, which was first proposed by Johann Daniel Titus in 1766, and given major prominence by Johann Bode in 1772, states that the radii R of the orbits of the planets as measured in Astronomical Units (AU) closely follow the law

$$R = 0.4 + 0.3 \times 2^n$$

where $n = -\infty, 0, 1, 2, 4, 5$ for the known planets Mercury, Venus, Earth, Mars, Jupiter and Saturn respectively. Uranus's discovery in 1781 at an orbital radius of 19.22 AU corresponded almost exactly with the predicted value for a planet with $n = 6$ (see Table 3.1).

This apparent endorsement of the Titus-Bode Law led to a frenzy of searches for the apparently missing planet with $n = 3$, which should have an orbit between Mars and Jupiter. Unfortunately the law gave no hint as to where in the sky the putative planet would be found except, presumably, somewhere reasonably close

n	Planet	Titus-Bode law	Observed orbit
$-\infty$	Mercury	0.4	0.39
0	Venus	0.7	0.72
1	Earth	1.0	1.00
2	Mars	1.6	1.52
3	–	2.8	–
4	Jupiter	5.2	5.20
5	Saturn	10.0	9.54
6	Uranus	19.6	19.22

Table 3.1 The Titus-Bode Law applied to solar system planets. Orbit sizes are expressed in astronomical units

to the ecliptic. The need for a large-scale search for the missing planet led to what is surely the first-ever example of "crowd-sourced" astronomy. In 1800 the German astronomer Franz Xavier von Zach organized an international collaboration of twenty-four astronomers who each undertook to focus their attention on a different fifteen-degree segment of the ecliptic plane. The group called themselves the "Lilienthal Detectives"; their most famous member was Heinrich Olbers, after whom the Olbers Dark-Sky Paradox is named.

The group had planned to invite the Sicilian astronomer Giuseppe Piazzi to join them, but in 1801, shortly before he received his invitation, he discovered a moving object in the sky, which he at first assumed to be a comet. Since the object moved behind the Sun soon after it was discovered it was not until a year later that its nearly circular orbit was confirmed and found to have a radius close to 2.8 AU—just as expected for a planet with $n = 3$ in the Titus-Bode Law. Piazzi named the object "Ceres" after the Roman goddess of agriculture.

Ceres's status as the unique missing planet was short-lived, however. First it was realized that Ceres is much smaller and fainter than any of the other planets. Secondly, the Lilienthal Detectives started to discover other objects in similar orbits around the Sun between Mars and Jupiter. Heinrich Olbers found Pallas in 1802, Karl Harding discovered Juno in 1804 and Olbers discovered Vesta in 1807. Because these objects appeared so small in a telescope, William Herschel suggested they be named "asteroids" after *aster*, the Latin word for a star. However many early nineteenth-century books can be found which refer to the solar system as having eleven planets: Mercury, Venus, Earth, Mars, Ceres, Pallas, Juno, Vesta, Jupiter, Saturn, and Uranus. Pluto is not the only "planet" which has suffered the indignity of demotion.

There then followed a hiatus of nearly 40 years in the discovery of asteroids, but six were discovered in the 1840s, and a further 300 by 1880. Asteroid discoveries came even faster after the development of photography because moving objects could be distinguished from stationary stars by the trails produced on long-exposure wide-field photographic plates. Nowadays vast numbers of new asteroids are discovered each week by digital surveys such as Pan-STARRS (see section 12.2).

3.2 Halley, Flamsteed and Navigation

The invention of the telescope did not stimulate any immediate efforts to re-map the sky, and nearly seventy years passed before the first telescope-based survey was published. One reason for this delay may be that performing accurate astrometry requires as much sophistication in the mounting of the telescope as in the quality of its optics.

Edmond Halley, after whom the famous comet is named, was born in 1656, fourteen years after Isaac Newton. In 1676 he travelled to St Helena, an island in the Atlantic Ocean at latitude 16° south, and set up several instruments including a 24-foot long telescope. Two years later he had completed his observations and returned to England where he published his *Catologus Stellarum Australium*, which contains positions and magnitudes for 341 stars. This volume broke ground in two ways: it was the first survey to be completed with a telescope and the first high-quality survey of the southern hemisphere.

The late 17th century was a splendid era for science in England; it encompassed the formation of the Royal Society of London in 1660, the publication of Isaac Newton's Principia in 1687 as well as the founding of the Royal Greenwich Observatory in 1675, under the direction of the first Astronomer Royal, John Flamsteed. A major motivation behind the funding of the observatory was the need to improve the reliability of nautical navigation, but it also yielded data relevant to some of the major questions of the day, including the orbits of comets and the nature of gravitation.

Flamsteed's main contribution at Greenwich was the compilation of a star catalog of unprecedented accuracy. Among his tools for making this catalog were a telescope firmly attached to a 10-foot radius brass quadrant on the side of a north-south wall (Figure 3.4), plus an accurate pendulum clock—a technological innovation that was not available to Tycho Brahe about a century earlier. He recorded the exact time that each star crossed the north-south meridian line from east to west, and measured its angle above the horizon off the precision scale on his quadrant. The version shown in Figure 3.4, built by Robert Hooke, was later replaced by a better version built to Edmond Halley's specification, and that instrument still exists at Greenwich Observatory.

Over the course of nearly 40 years Flamsteed made about 20,000 observations accurate to about ten seconds of arc. His entire catalog of 2935 stars was not published until 1725—several years after his death. The delay, which was mainly the result of his meticulousness, led to a falling out with Isaac Newton, who attempted to steal and prematurely publish Flamsteed's data for the purposes of testing one of his theories.

Quadrans Muralis merid: 10 pedum Rad:.

Fig. 3.4 10-foot mural quadrant made by Robert Hooke for the Royal Observatory, Greenwich, about 1676

Flamsteed's star catalog, one page of which is shown in Figure 3.5, was one of the first to use the equatorial coordinate system—the one that uses right ascension and declination. Earlier catalogs, from Hipparchus up to Halley, listed star positions only in ecliptic coordinates. The essential difference between these systems is that equatorial coordinates are tied to the spin axis of the Earth, while ecliptic coordinates are tied to the orbit of the Earth around the Sun. They differ because the Earth's spin axis differs from the axis of its orbit by an angle of about 23.5°. The relative advantages of these two systems is discussed in Appendix A.7.

STELLARUM INERRANTIUM.

In Constellatione TAURI.

ORDO		STELLARUM Denominatio.	Bayer. Cha.	Ascensio Recta	Distantia à Polo B.	Longitudo.	Latitudo.	Varia. Asc. R.	Varia. D à P.	Magnitudo.
Ptol.	Tycho			o ′ ″	o ′ ″	s o ′ ″	o ′ ″	′ ″	′ ″	
29	29	Quæ Borea sequentis lateris Quadrilateri in Cervice.	φ	60 18 0	63 26 15	II 3 32 59	5 46 12 B	65 18	12 3	5
				60 18 10	69 39 5	2 19 18	0 19 23 A	62 39	12 2	7
11	11	De quinque in facie dictis Suculæ est in Naribus	γ	60 32 40	75 9 55	1 27 34	5 46 22 A	60 30	11 57	3
				60 32 45	74 7 55	1 39 52	4 45 35 A	60 55	11 57	7
				60 35 10	69 0 55	2 42 21	0 15 0 B	62 34	11 57	7
				60 37 40	76 45 0	1 13 31	7 20 32 A	59 52	11 56	6.7
			b	60 45 10	75 41 45	1 33 12	6 19 57 A	60 17	11 54	7
28	28	Australis sequentis lateris Quadrilateri in Cervice	χ	60 56 30	65 9 0	3 46 56	3 58 41 B	64 36	11 48	5
				61 3 20	76 41 30	1 38 53	7 22 0 A	59 57	11 48	7
12	12	Inter illam in Naribus (γ) & Oculum Bor. } 1 ad	δ	61 16 10	73 13 45	2 31 27	4 0 34 A	61 18	11 38	4
				61 20 45	66 27 45	3 53 21	2 37 6 B	64 6	11 38	7
				61 25 0	73 59 0	2 31 24	4 44 58 A	61 1	11 39	6
		2 ad	δ	61 34 20	73 19 5	2 47 13	4 9 4 A	61 16	11 35	4
23	23	Duarum propinquarum in Aure Borea, Austrina }	κ	61 44 10	68 27 35	3 51 53	0 35 21 B	63 15	11 29	5
	35		r	61 44 40	31 18 15	1 24 57	12 1 21 A	58 14	11 29	5
		2 ad	κ	61 45 0	68 33 10	3 51 37	0 29 46 B	63 13	11 31	5
		3 ad	δ	61 54 0	72 49 20	3 11 42	3 43 27 A	61 30	11 29	6
22	22	Duarum propinquarum in Aure borea, Borealis } 1 ad	υ	61 57 10	67 56 0	4 9 42	1 4 6 B	63 29	11 25	5
				61 59 10	74 48 55	2 54 1	5 41 50 A	60 43	11 24	7
				62 11 30	75 7 55	3 2 10	6 2 44 A	60 36	11 21	7
22		2 ad	υ	62 12 0	67 44 45	4 25 18	1 12 36 B	63 36	11 20	6
	37	Præcedens trium. infra Suculas.	π	62 16 40	76 2 5	2 56 57	6 56 53 A	60 14	11 17	5
15	15	In Oculo Boreo	ε	62 38 0	71 32 45	4 7 11	2 35 58 A	62 3	11 14	3·4
				62 41 30	74 22 45	3 39 12	5 23 43 A	60 54	11 9	7
				62 42 40	75 59 35	3 22 25	6 59 1 A	60 16	11 10	7
13	13	Inter 11ᵐ (γ) & Austrinum Oculum } 1 ad	θ	62 43 10	74 46 25	3 36 25	5 47 16 A	60 45	11 9	5
	36	In sinistro Genu	b	62 44 40	74 51 55	3 36 51	5 52 55 A	60 44	11 9	5
				62 52 0	77 41 5	3 12 31	8 40 32 A	59 37	11 5	5
				63 7 10	75 4 30	3 56 0	6 9 18 A	60 42	10 58	7
				63 14 40	75 1 5	4 3 47	6 7 13 A	60 51	11 0	7
				63 14 40	75 36 30	3 57 22	6 42 4 A	60 29	10 56	7
				63 17 30	76 59 45	3 44 57	8 4 25 A	59 56	10 57	7
				63 23 20	75 36 25	4 5 43	6 43 28 A	60 28	10 56	7
				63 32 30	74 51 30	4 22 35	6 0 53 A	60 46	10 52	7
	38	Media trium infra Suculas	ρ	64 4 0	75 51 15	4 42 7	7 5 6 A	60 26	10 41	5
14	14	Splendida in Austrino Oculo Palilicium, Aldebaran }	α	64 32 20	74 9 40	5 27 0	5 29 49 A	61 7	10 28	1
10	10	In sinistro Cubito	d	64 39 30	80 31 15	4 27 10	11 46 45 A	58 37	10 25	5
				65 6 0	74 38 5	5 54 15	6 3 20 A	60 57	10 16	7
9	9	In sinistro Genu, 1 ad	c	65 12 30	78 9 15	5 24 30	9 32 32 A	59 33	10 12	5
	39	Ultima trium infra Suculas } 1 ad	σ	65 22 0	74 51 40	6 7 14	6 19 19 A	60 51	10 10	6
		2 ad	σ	65 23 30	74 44 35	6 9 52	6 12 35 A	60 55	10 9	6
9	9	In sinistro Genu 2 ad	c	65 41 40	78 27 25	5 49 58	9 55 14 A	59 26	10 2	5
20	20	In Eductione Cornu Borei	τ	65 55 10	67 40 50	7 49 20	0 40 23 B	63 54	9 58	5
				66 20 10	66 32 30	8 23 8	1 44 11 B	64 25	9 48	6.7
				68 0 15	74 35 55	8 41 27	6 28 0 A	61 4	9 8	6
		In Eductione Cornu Austri, & in Aure	i	68 19 0	71 43 55	II 9 24 58	3 40 35 A	62 20	9 3	6†

Fig. 3.5 One page from Flamsteed's 1725 star catalog

3.3 Messier, Herschel, and Nebulae

When looked at directly though a telescope, all stars except for the Sun appear as points of light; even the best modern telescopes have to be fitted with special equipment in order to measure or resolve their angular diameters. Even nearby supergiant stars such as Betelgeuse are only about a twentieth of an arcsecond in angular diameter as viewed from the Earth—ten thousand times narrower than the full moon. Hence any object that appears fuzzy to the naked eye or through a conventional telescope must be something other than a single star.

Several pre-telescopic astronomers noticed a small number of objects which appeared to be fuzzy or extended. Ptolemy lists five of them, including the star cluster we now refer to as h and χ Persei, and Ulugh Beg, as mentioned in section 2.3, drew attention to both the Andromeda galaxy and the Large Magellanic Cloud. However no astronomer in the pre-telescope era seems to have given nebulae the attention they deserve.

Within a year of Galileo's first use of a telescope in 1610, however, the Frenchman Nicolas-Claude de Peiresc had discovered the Orion Nebula. Further discoveries were made as telescopes improved over the next century, but the progress was slow until the astronomer Charles Messier published his famous catalog of nebulae in 1771 based on his observations with a 10-cm diameter telescope.

Messier's catalog is one of the most splendid paradoxes in the history of science. It is a list of about 100 fuzzy objects that could be safely ignored by anyone interested in hunting for comets, which was his own enthusiasm. What he failed to appreciate is that his list includes the prime examples of at each least ten other celestial phenomena that many astronomers would consider to be of considerably greater astrophysical importance than comets. Examples are shown in Figure 3.6; they include six different kinds of objects found within the Milky Way, and four different kinds of galaxy.

At about the same time that Messier was making his survey, the German-born, but England-resident, astronomer William Herschel (1738–1822) was building a series of ever more powerful telescopes. For the first thirty years of his life Herschel earned his living as a musician, composing over 20 symphonies, but astronomy was his ardent hobby and by 1781 he had built a 7-foot long reflecting telescope with a 6.2-inch diameter mirror that was better than anything else in England (Figure 3.7). Like almost all reflecting telescopes until the late 19th century, the mirror was made of speculum metal, an alloy of tin and lead that could be shaped and polished to make a precision mirror. It was with this instrument, and a subsequent 20-foot long instrument with an 8.5-inch mirror, that he made his historic sky surveys.

Herschel's initial strategy was to examine known stars, looking for evidence of binary companions or nebulosity, but the publication of Messier's catalog encouraged him to search more widely, making sweeps of the heavens in the search for new phenomena. He ultimately discovered over two thousand nebulous objects which he published in several catalogues between 1786 and 1802. He sorted them into eight classes according to their appearance. Many of these we now know to be galaxies as opposed to star clusters and gaseous nebulae in our Milky Way, but

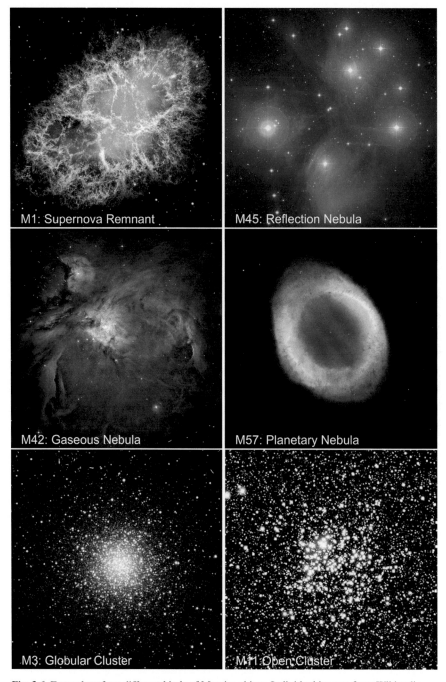

Fig. 3.6 Examples of ten different kinds of Messier object. Individual images from Wikipedia

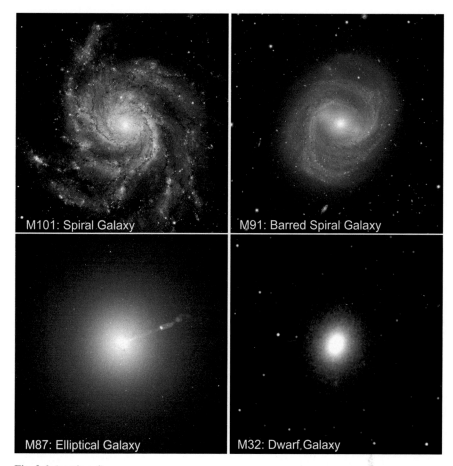

Fig. 3.6 (continued)

the nature of galaxies was not recognized until the early part of the 20th century. William Herschel's work, as extended by his sister Caroline and his son John, was eventually published as the General Catalog of Nebulae and Clusters in 1864 and contained about 5,000 entries.

Herschel's work on cataloging nebulae was extended by the Danish-Irish astronomer John Dreyer who published his New General Catalog in 1888 and its supplements, the Index Catalogs, in 1895 and 1907. These catalogs are the origin of the widely-used NGC and IC designations for galaxies and nebulae, and together contain some 13,000 objects. These catalogs are a compendium of data from many different astronomers and telescopes, so strictly speaking they are not in themselves true surveys.

Fig. 3.7 Replica of William Herschel's 7-foot telescope. Image credit: Mike Young

3.4 Variable Stars and Photometry

Before the invention of the telescope all stars were believed to shine with constant brightness, with the exception of an occasional nova such as the supernova observed by Tycho Brahe. The first astronomer to recognize periodic variability in a star was Johannes Holwarda in 1638, who found that the star Omicron Ceti had a period of approximately 11 months. The star was later given the special name Mira and has become the prototype for an important class of long-period variable stars that now number more than 6,000.

More variable stars were discovered in the 1780s by the English astronomers John Goodricke and Edward Pigott who, with some input from Herschel, were eventually able to draw up a list of 39 variables. They included the eclipsing binary stars β Persei (Algol) and β Lyrae as well as δ Cephei, the prototype Cepheid variable star.

Early studies of variability were generally based on the fact that the eye is very good at deciding which of two stars is the brighter. An astronomer looking for variability would try to match the star under study with one of its nearby neighbors.

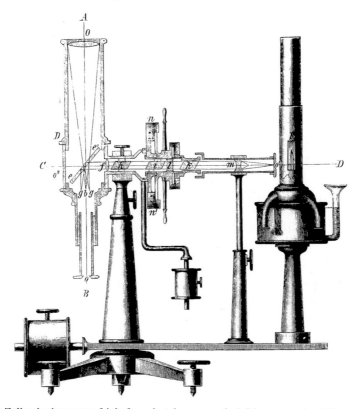

Fig. 3.8 Zollner's photometer. Light from the telescope on the left is compared with light produced by a flame on the right. The polarizing Nicol prisms can be seen in the horizontal tube

If he got a better match by using a different star next time he looked at it he would log it as a suspected variable star and monitor it for periodicity. Remarkably, Herschel was able to estimate brightness changes with a precision of better than 0.15 magnitudes using this method.

The next major step forward in the study of stellar brightnesses and variability came in the mid nineteenth century with the development of the comparative photometer, first by Carl von Steinheil (1801–1879) and later by Friedrich Zollner (1834–1882). In Zollner's photometer (Figure 3.8) the astronomer matched the light from the star under study with the light from a kerosene lamp by adjusting a filter in front of the lamp until the images matched in brightness. The filter consisted of two Nicol prisms which each acted like the polarizing lenses of a modern pair of sunglasses. As the astronomer rotated one of the prisms with respect to the other the amount of light passing through them ranged from 50 % to zero, and could be calculated from the rotation angle. The Zollner photometer was widely used during the second half of the nineteenth century.

The Harvard astronomer Edward Pickering, whom we will meet again in the next chapter, built an improved version of the photometer and in the period 1879–1882 produced a catalog of the brightnesses of over 4,000 stars. The development of the photometer also led to a major increase in the number of known variable stars; by 1884 there were nearly 200 catalogued.

3.5 Moving Stars

As the positions of stars could be measured with ever-greater precision, astronomers started to see if they could detect changes in their positions over a period of time. These searches led to three major discoveries: proper motions, binary stars, and parallaxes. These effects are all much smaller than precession, which had been discovered in ancient times (see section 2.2).

Proper Motions The term proper motion refers to a steady detectable change in a star's position in the sky relative to other stars. The first proper motions to be measured were by Edmond Halley who noticed that Flamsteed's positions for Sirius, Arcturus and Aldebaran differed by over half a degree from those measured by Hipparchus nearly 2000 years earlier. This discovery was followed up by several people including Tobias Mayer of Gottingen who measured proper motions of 80 stars by comparing his own measurements with those made by Nicolas Lacaille about 50 years earlier. Meyer realized that if the solar system itself was in motion among the stars, one might see a systematic pattern in the proper motion. He failed to find any such evidence in his data, but in 1783 William Herschel found exactly such an effect when he compared his measurements with those made earlier by Nevil Maskelyne; he determined that the solar system is moving towards the Hercules constellation, but since he did not know the distance to any of his stars, he could not say what the velocity was.

Surveys for stellar proper motions became more common after the development of astrophotography; in 1919 Max Wolf at the Heidelberg observatory published a catalog of over a thousand proper motions.

Double Stars Very few double stars were known before the invention of the telescope, but by 1767 so many were known that an English clergyman named John Mitchell was able to argue statistically that the majority of double stars must be physical pairs rather than chance encounters. While William Herschel was surveying the sky for nebulae he was also on the lookout for double stars and multiple stars and by 1784 he had discovered about 700 systems which he published in two catalogs. By 1802 he had remeasured many of them and found changes that convinced him that at least some of them were bound together by their mutual gravitational attraction. We usually refer to these systems as binary stars.

Nowadays astronomers interested in double stars make use of the *Washington Double Star Catalog* published online by the United States Naval Observatory, which lists over 100,000 double star systems. Almost everything astronomers know about the masses of stars and black holes comes from the study of binary systems.

Parallax As the Earth moves around the Sun, our view of our three-dimensional universe alters slightly; nearby stars appear to move in tiny ellipses against the relatively fixed background of distant stars. If we can measure the size of one of these ellipses (in fractions of a degree) we can calculate the distance to that star, so long as we know the size of the Earth's orbit around the Sun. This effect is called "Stellar Parallax." The idea behind parallax was clear to the ancient Greeks who looked for changes in the sky they could attribute to the motion of the Earth round the Sun. The fact that they did not see any such effect was evidence to them that the Earth did not orbit the Sun and that the Earth was the center of the universe. What the Greeks did not appreciate is that the distances to all the stars other than the Sun are so large that their parallaxes are always less than one second of arc—far too small to be detected by the naked eye, or even during the first two centuries of telescope observation. Of the dynamical effects discussed in this section, parallax is by far the most difficult to measure.

The first stellar parallax was measured by Friedrich Bessel in 1836, who measured the distance to the star 61 Cygni. Further results came slowly and only about 60 parallaxes had been measured by the end of the 19th century. Results came faster during the photographic era of the 20th century, but the largest leap came in 1989 with the launch of the Hipparcos satellite which was able to measure parallax distances to stars up to 1600 light years away (see Chapter 11).

3.6 The BD Catalog

The ultimate star catalog of the pre-photographic era was the *Bonner Durchmusterung* usually referred to as the BD catalog. This catalog, the first part of which was published in 1859, lists some 320,000 stars brighter than 10th magnitude. Work on the catalog was led by the German astronomer Freidrich Argelander (1779–1875) at the Bonn observatory which he had earlier founded. Figure 3.9 shows the telescope that he and his assistants used. Although its objective lens had a diameter of only 7.5 cm and its magnifying power was only 10, its simplicity made it a very efficient tool for its purpose.

The 1859 catalog and its 1886 extension covered stars north of declination $-23°$. It was supplemented in 1892 by the 580,000 star *Cordoba Durchmusterung* catalog of stars between $-22°$ and $-90°$ from Cordova observatory in Argentina. The final catalog in the BD series, the Cape Photographic Durchmusterung (450,000 stars, 1896) covered the whole of the southern hemisphere from South Africa and made use of photography.

Fig. 3.9 Argelander's telescope at the Bonn Observatory. The large wheels carried the scales with which a star's position could be accurately measured. Image credit: Michael Geffert

The numbering system of the BD catalog is still widely used to identify stars: Betelgeuse, for example is numbered BD+07 1055, meaning it is the 1055th star in the declination range $+7°$ to $+8°$, starting from zero right ascension. This catalog became the basis for the HD and SAO star catalogs published in the 20th century. These catalogs contain roughly the same number of stars as the BD catalog, and represent an upper limit to the sizes of star catalogs published on paper. Their recent successors, the Hipparcos and Gaia space missions, which are described in Chapter 11, exist only in electronic form.

Chapter 4
The Photography Era

Although experiments with light-sensitive chemicals began in the early 19th century, it was not until the late 1830s that Louis Daguerre in France and Henry Fox Talbot in England found ways to indefinitely preserve an image on either glass or paper. Early emulsions required very long exposures, but by 1850 the daguerrotype process had been sufficiently improved for an image of the star Vega to be obtained using the 15-inch refractor at Harvard College Observatory.

The light sensitivity of photographic plates improved rapidly, leading to a breakthrough in 1883, when the London-based amateur astronomer Andrew Common took a 60-minute exposure with his 36-inch refracting telescope that for the first time showed stars that were too faint to be seen with the naked eye (Figure 4.1). The photographic era of astronomy had begun. It lasted until about the 1990s, by which time digital imaging had taken over almost completely, as we shall describe in Chapter 12.

Astrophotography was a complicated process involving thin glass plates that were covered in a light-sensitive emulsion containing compounds of the element silver. Flexible plastic photographic film, although cheaper, was rarely used by professional astronomers, since it cannot be relied on to maintain its shape indefinitely. Each exposure—which might last anywhere from minutes to hours—was made on a new plate that had to be carefully inserted into the telescope's focal plane. The plates had to be handled in complete darkness both before and after the exposure, and were then developed in a darkroom that was usually inside the observatory building itself. The developing process involved immersing the plates in a series of chemical solutions for pre-determined periods of time, and then allowing them to dry. Depending on the project, and on whether an assistant was available, an astronomer might spend the whole night oscillating between the telescope and the darkroom.

The chemistry of the photographic process means that images on developed plates are negatives, with stars appearing as black spots against a white sky background. Astronomers got so used to dealing with negative images that they frequently appeared that way in scientific publications. Positive images, such as that in Figure 4.1, required a second photographic process.

© Springer International Publishing Switzerland 2016
G. Wynn-Williams, *Surveying the Skies*, Astronomers' Universe,
DOI 10.1007/978-3-319-28510-8_4

Fig. 4.1 Andrew Common's 1883 photograph of the Orion Nebula

4.1 Carte du Ciel

Very soon after Common's photograph was published, a major project was initiated to make a photographic survey of the whole sky. In 1887 the director of the Paris Observatory, Amédée Mouchez convened a congress of 50 astronomers from 20 observatories in both hemispheres (Figure 4.2). They agreed to share the task of surveying the sky to 14th magnitude. The observatories equipped themselves with similar 13-inch diameter, 11-foot long astrographic telescopes (Figure 4.3), and proceeded to photograph the sky in a series of $2° \times 2°$ fields on 13 cm square glass plates. Each observatory was responsible for a particular range of declinations.

The project had two goals. The first was to produce a reference catalog of star positions down to magnitude 11.5; this was called the "Astrographic Catalogue," to be published as a list of positions. Exposures for this part of the project were typically six minutes. Stars on each developed plate were measured manually using about a dozen reference stars on each plate. This effort yielded a staggering 254 individual volumes of star positions, which were published independently by the various observatories involved. One effect of this diversification is that the format of the catalogs differ. Some declination ranges have star positions in equatorial coordinates accurate to better than 1 second of arc. Others list only the x and y

Fig. 4.2 Astronomers at the 1887 congress at Paris Observatory

coordinates measured from a given plate center, and leave the reader to calculate the equatorial position using a provided formula. In all some 4.6 million stars were observed and catalogued.

The second part of the project, the Carte du Ciel itself, involved taking deeper exposures with the same telescopes, reaching down to about 14th magnitude. The plan was to print and distribute enlarged reproductions of these images using engraved copper plates. Parts of this project were completed and published, but a number of observatories did not complete, and in some cases did not even start this deep survey. About half the sky was covered: most of the plates still exist in the archives of their home observatories.

The Carte du Ciel project dragged on far longer than anticipated when it was set up. The last images were taken in 1950 and the last volume of the Astrographic Catalogue was published in 1964. Until recently the project was regarded as something of a glorious failure: it had absorbed the energies of many European astronomers for far longer than originally planned at a time when astronomers in the USA were turning their attention to the new worlds of spectroscopy and astrophysics. But about a century after it was first planned, the Carte due Ciel finally showed its worth. All the volumes of the Astrographic Catalogue were converted to digital form at the Sternberg Astronomical Institute in Moscow. The equatorial positions of all 4.6 million stars in the catalog were then re-computed at the USA Naval Observatory in Washington using reference stars from the Hipparcos astrometry satellite (see Chapter 11). A comparison of the precise star positions as they were about a century ago with their positions in the 1990s as measured by the Hipparcos satellite has now given us the proper motions of 2.5 million stars: an invaluable tool for studying the clustering of stars and the dynamics of the Milky Way Galaxy.

Fig. 4.3 One of the telescopes used to make the Carte du Ciel—the 13-inch astrograph at Greenwich Observatory, London in about 1900. Image credit: Graham Dolan

4.2 Harvard Plate Collection

At about the same time as the Carte du Ciel was being started, a quite different kind of photographic survey was initiated at Harvard College Observatory. The purpose of this project was not to produce a catalog or an atlas, but to repeatedly photograph the sky in order to support the rapidly-developing science of variable stars. From 1885 to 1993, with a gap between 1953 and 1968, regular photographs of the whole northern hemisphere sky were taken from Harvard, while corresponding images of the southern sky were obtained at its sister observatories in Peru and South Africa.

In all, some 500,000 glass plates were exposed, almost all of which are stored in an elaborate temperature and vibration-controlled vault at Cambridge, Massachussets (Figure 4.4). Most parts of the sky were imaged several hundred times and some parts more than a thousand times during the 100-year lifespan of the project. On most plates any star brighter than about 15th magnitude can be measured; for the best plates, stars as faint as 18th magnitude can be seen.

Fig. 4.4 A small part of the Harvard Plate Collection. Image credit: Harvard College Observatory

An astronomer with an interest in a particular variable star can apply for permission to visit the vault in Cambridge and study the images of the star on the relevant time-series of plates, using cameras or eyepieces attached to microscope-like viewers. The brightness of a star on a particular date is determined by comparing the size of its image with those of nearby standard stars on the same plate. The astronomers who make use of this opportunity include both professionals and keen amateurs.

The acquisition of new images ended in 1993; the advent of digital imaging was leading to the decline in the availability of glass photographic plates. But in 2004 the project started a new lease of life when it was announced that a specially-designed digital scanner had been designed and installed in the vault, and that it would be used to make digital images of all the glass plates in the collection. The scanner in question is faster than any predecessor: it can scan two 8-inch by 10-inch plates in just 90 seconds. Even so, it will take several years to complete the task of converting all the important information in all the stored plates into a form that can be accessed by anyone with an internet connection.

4.3 Schmidt Camera Surveys

Perhaps the most widely used of all the photographic surveys is the Palomar Observatory Sky Survey (POSS), which was carried out between 1949 and 1958 using the 48-inch (1.2-m) Schmidt Telescope on Mount Palomar, California (Figure 4.5). Schmidt telescopes, of which this was the largest ever built, are specifically optimized for wide-field astrophotography rather than for, say, spectroscopy; for this reason they are sometimes referred to as cameras rather than telescopes. The Palomar telescope recorded its images on 14-inch square glass plates that each covered a $6° \times 6°$ field of view on the sky.

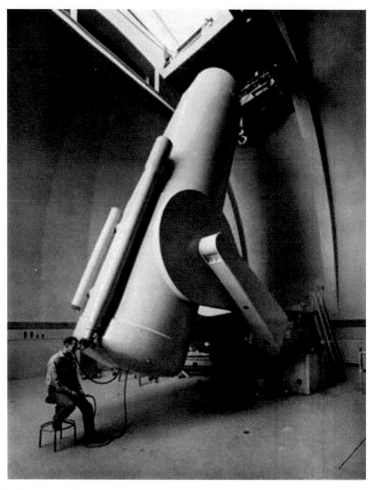

Fig. 4.5 The 48-inch Schmidt Telescope at Mount Palomar. Image credit: Palomar Observatory and Caltech

For the original survey, two exposures were taken in quick succession, one using a blue filter and one using a red filter; by comparing the images of an object in these two exposures an astronomer could get an estimate of the color of a star, which is a guide to its surface temperature. The original survey included 937 pairs of exposures covering the whole sky north of declination $-27°$. The limiting magnitude of the survey was about 22. The glass plates were used to make 14- by 17-inch square negative paper photographic prints (Figure 4.6) which were sold to interested researchers round the world. The whole set of paper prints could be purchased and stored in a couple of filing cabinets.

Several years after the original Palomar survey was completed a complementary survey was conducted of the sky south of declination $-17°$. The work was shared between the European Southern Observatory's 1-m Schmidt camera in Chile, which exposed the red plates, and the 1.2-m United Kingdom Schmidt Telescope in Australia which produced the blue plates. The southern survey, which consists of 606 pairs of plates, was completed in 1990.

The greatest value of the Palomar Sky Survey has probably been its impact on other branches of astronomy such as radio astronomy and infrared astronomy, whose early pioneers usually had backgrounds in physics rather than optical astronomy. When these astronomers discovered new sources of radiation it was to the Palomar Sky Survey that they usually turned first. It was easy to use because all the prints were at the same scale and sensitivity, and it did not require expensive equipment to maintain and study.

Fig. 4.6 A pair of Palomar sky survey prints. Each print covers a sky area of 6° by 6°. The very bright stars at the bottom right are Orion's Belt

In the early 1980s, the Palomar Schmidt Telescope was upgraded and used to make a second survey, "POSS II". Both this survey and the original "POSS I" have now been digitized and made freely available on the web. The telescope has since been renamed the Samuel Oschin Telescope and has been adapted to use digital cameras rather than photographic plates.

4.4 Harvard Spectroscopic Survey

Except perhaps for photography, the most important astronomical development in the nineteenth century was the introduction of spectroscopy, first of the Sun, and then of stars. Dark lines in the Sun's spectrum were first noticed by the chemist William Wollaston in 1802 and subsequently catalogued by Joseph von Fraunhofer, the inventor of the spectroscope. In 1832 Fraunhofer went on to discover similar absorption lines in the spectra of a number of bright stars, including Betelgeuse and Sirius.

It was not until 1859 that the origins of these lines became understood, when Gustav Kirchhoff of Heidelberg University recognized that a pair of strong yellow absorption lines seen in many stars were identical with absorption lines he saw in his laboratory when he passed white light through a flame contaminated with a sodium compound such as common salt. Within five years eight more elements were identified, mainly by William Huggins in London.

Huggins soon made two other major spectroscopic discoveries. In 1864 he noted that the spectra of nebulae were of two quite different types: those of planetary nebulae showed spectral lines in emission, while nebulae that we would now identify as galaxies have absorption-line spectra, much more like those in stars. Then in 1868 he was the first person to measure a Doppler shift of light waves, noting that the wavelength of one of the spectral lines of hydrogen in Sirius was about 0.01 % longer than that seen from the stationary standard light source.

The first spectroscopic survey of stars was made in 1866 by the Italian astronomer Angelo Secchi, His catalog of stellar spectra contained about 220 stars that he had observed visually. He used differences in the relative strengths of the absorption lines to classify the stars into four groups, and was one of the first astronomers to stress the idea that the Sun itself is a star.

The first person to apply photography to the recording of stellar spectra was Henry Draper, an American doctor and amateur astronomer, who obtained the first photographic spectrum of a star (Vega) in 1872. He subsequently obtained spectra of a hundred more stars at his private New York observatory before his untimely death in 1882.

Draper's work was followed up by Edward Pickering at Harvard College Observatory. Pickering set out to obtain the spectra of as many stars as possible by placing a large, thin prism in front of the lens of the telescope, a technique known as objective prism spectroscopy. In this way he could obtain the spectra of up to 200 stars on one plate (Figure 4.7) and classify them into groups based on the

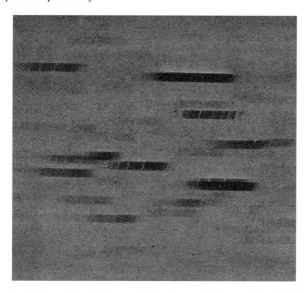

Fig. 4.7 Part of a plate used for the Harvard Spectroscopic Survey. The light from individual stars has been split into small spectra which are used to classify each star according to its stellar temperature. Image credit: Harvard College Observatory

Fig. 4.8 Edward Pickering of Harvard College Observatory with his staff of "computers" in 1913

absorption lines seen in each star's spectrum. He recruited a large staff of women "computers" to analyze the observations (Figure 4.8), several of whom went on to become prominent astronomers in their own right. One of them, Annie Jump Cannon, was primarily responsible for the OBAFGKM stellar classification scheme that is still in wide use today. We now know that these spectral classifications refer to the surface temperature of a star, from the hot (type O) to cool (type M), but this connection could not be properly addressed until the 1930s, when the structure of atoms became clarified by the new science of quantum mechanics.

The resulting final catalog of 220,000 stars classified according to spectral type was published between 1918 and 1924 as the Henry Draper Catalog or HD catalog for short. It was so named because the project had been generously funded by Draper's widow, Mary. The 9-volume catalog covers both hemispheres and is complete down to about 9th magnitude. Extensions to the catalog in the 1930s and 1940s brought the total number of stars classified to nearly 360,000.

Chapter 5
Radio Surveys

We have covered the first three eras in astronomy in one chapter each. The fourth era—what one might call the Electromagnetic Era—will require six chapters. This is the era in which astronomers started to collect and study celestial radiation at wavelengths far outside the range of visible light. The dawn of this era came in the 1930s with the development of radio astronomy, but because many kinds of electromagnetic radiation are blocked by the Earth's atmosphere, most of the other new disciplines had to wait for the space age, which started in the 1960s. In this chapter we will discuss radioastronomy: the other wavebands will be covered in Chapters 6 to 10.

Radio waves accessible to astronomers have wavelengths between about 30 m and 1 mm (see Figure 5.1). These limits are set by the Earth's atmosphere: wavelengths longer than about 30 m tend to get distorted and blocked by free electrons high in the Earth's ionosphere, while those shorter than around 1 mm are absorbed by water vapor in the lowest few kilometers of the atmosphere. Astronomy at wavelengths of around 1 mm is dominated by the phenomenon of the Cosmic Microwave Background (CMB); this subject is therefore treated separately in Chapter 7.

At most radio wavelengths, astronomical observations are impossible because of interference from man-made communication and radar signals. Fortunately, there are international agreements that keep about 1 % of the radio spectrum free of man-made signals, and it is in about a dozen of these "windows" that most radio astronomy is done. The most important of these reserved bands is at 21 cm wavelength, where neutral hydrogen atoms have a unique spectral line (section 5.6).

5.1 Jansky and Reber

The first person to unambiguously detect extraterrestrial radio signals was Karl Jansky, an engineer working for the Bell Telephone Company in New Jersey, during the early 1930s. Tasked with tracking down signals that might interfere with the blossoming telecommunications industry, he built a special rotating antenna which

© Springer International Publishing Switzerland 2016
G. Wynn-Williams, *Surveying the Skies*, Astronomers' Universe,
DOI 10.1007/978-3-319-28510-8_5

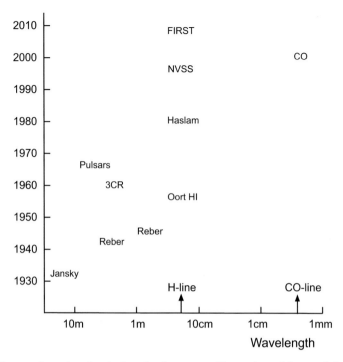

Fig. 5.1 Dates and wavelengths of selected radio surveys. Observations of the cosmic background radiation are shown in Figure 7.1 rather than here

could scan the horizon at a wavelength of 14.5 m (Figure 5.2). He noticed that the direction from which the strongest noise signal came changed from day to day. It was correlated neither with the direction of the Sun nor any fixed point on Earth. After a long series of measurements in 1932, he concluded that the emission came predominantly from the Milky Way rather than from any single point in the sky. He remarked on the fact that the Sun was too faint to be detected by his antenna and speculated that the galactic emission might therefore be generated in the interstellar medium rather than by the stars. The era of radio astronomy had begun.

Jansky's work was received with respect, but was largely ignored by the astronomers of the day, who showed little interest in photons that could not be recorded on a photographic plate. It was left to another radio engineer, Grote Reber, to follow up Jansky's discovery; from 1937 until 1943 he was the world's only radio astronomer.

Fig. 5.2 Karl Jansky and his antenna. Image credit: Alcatel-Lucent USA Inc

Working entirely in his spare time, Reber designed and built a 10-m diameter parabolic dish reflector in the back yard of his house in a Chicago suburb (Figure 5.3), and fitted it with a series of cutting-edge radio receivers. With this instrument he mapped the whole northern sky at 1.9 m wavelength and much of it at 0.6 m (Figure 5.4). He confirmed that the radio emission from the sky was strongly correlated with the location of the Milky Way, being strongest towards the center of the galaxy in Sagittarius. The correlation was not perfect, however, and his maps show hints of the very powerful compact radio sources that would later become known as Cassiopeia A and Cygnus A.

Reber chose to make his observations at shorter wavelengths than Jansky for two reasons:

First, he knew that the laws of physics set a limit to the detail that he could see with his telescope. As discussed in Appendix A.4, the resolving power of a radio telescope, often loosely referred to as the beamwidth, depends on the ratio of the wavelength being observed to the diameter of the antenna; the shorter the wavelength, the finer the detail. This effect can be seen in Figure 5.4; the maps made at a wavelength of 0.6 m shows finer detail than the ones made at 1.9 m. If we apply the formula given in the appendix to Reber's 10-m diameter telescope we can derive a beamwidth of 12° for his 1.6-m observations, which corresponds reasonably well with the finest detail discernible in the upper left map.

Fig. 5.3 Grote Reber's antenna. Image credit: Wikipedia

Second, he made the reasonable guess that the radio emission from the Milky Way arises as a result of the random thermal motions of charged particles in a hot gas of some kind. The laws of physics imply that if this were the case the strength of the radio emission ought to be stronger at short wavelengths than long wavelengths. He was therefore very surprised to discover that the radio emission actually decreases in strength as one moves to shorter wavelengths. This decrease meant that the emission was caused by some process other than radiation from a hot gas, and explaining its origin became a astrophysical challenge. It was only in 1950 that it was realized that most of the radio emission from the Milky Way galaxy is caused by synchrotron radiation, which is generated by cosmic ray electrons spiraling around magnetic field lines at speeds close to that of light.

It is interesting to compare Reber's map of the Galaxy with one made about 35 years later using much larger and more sensitive radio telescopes, Figure 5.5 was published in 1981 by Glyn Haslam and his colleagues using data combined from the 100-m diameter telescope in Bonn (Germany), the 75-m telescope in Jodrell Bank (UK) and the 64-m telescope in Parkes (Australia) (Figure 5.6). The finest details

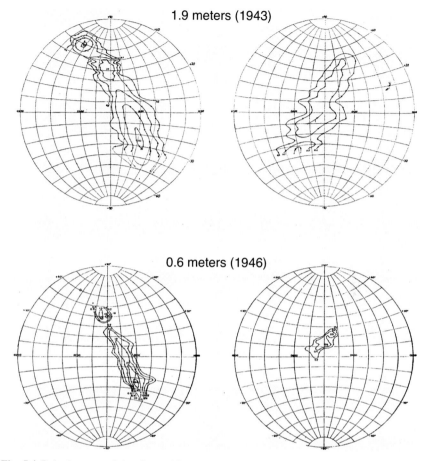

Fig. 5.4 Reber's maps of the sky at 1.9 m wavelength and 0.6 m wavelength. The maps are in equatorial coordinates: the center of the Milky Way galaxy coincides with the strongest peak of emission at declination of around $-25°$. Image credit: ©American Astronomical Society, from Ap.J. 100 279 (1944), and ©Institute of Electronic and Electronic Engineers, from Proc I.R.E. 36, 1215 (1948)

that can be seen are roughly one degree across. This image is displayed in galactic coordinates (see Appendix A7) and confirms that the bulk of the radio emission from the sky arises from near the plane of the Milky Way. But the image also shows several discrete radio sources both in the plane of the Milky Way and far from it. Understanding the nature of these objects became a prime goal of radio astronomers in the 1940s and early 1950s, the period to which we will now return.

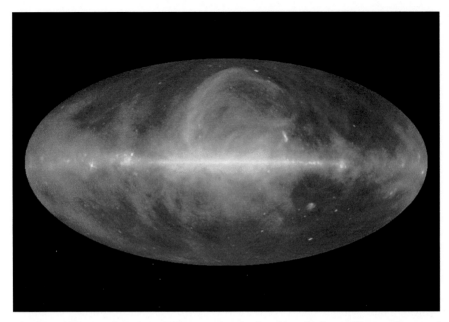

Fig. 5.5 A 1981 map of the sky at 0.75-m wavelength, plotted in galactic coordinates and centered on the nucleus of the Milky Way galaxy. Most of the emission is generated by the synchrotron process. Image credit: Max Planck Institute for Radio Astronomy, Bonn, based on the survey by Haslam et al. AASS 47, 1 (1982)

Fig. 5.6 The 64-m diameter Parkes radio telescope in Australia, built in 1961. Image credit: Ian Sutton

5.2 Post World War II Radio Astronomy

The development of radar during the Second World War honed the skills of a number of young scientists, particularly in the UK and Australia, providing them with the skills and enthusiasm to experiment with the new science of radio astronomy. Much of their early post-war work involved the study of the Sun, from which sporadic radio emission had been detected during the war. However, it soon became clear that the greatest mysteries lay beyond the solar system and that solving them would require instruments that could discern much finer detail than was possible with existing radio telescopes.

Broadly speaking, radio astronomers divided into two camps. While some of them advocated the construction of very large parabolic antennas, such as the 64 m diameter dish at Parkes, Australia (Figure 5.6), others realized that a great deal more information could be obtained by employing the principle of interferometry. In its simplest form a radio interferometer consists of two antennas pointing at the same piece of sky but separated on the ground by a hundred meters or more. The signals from the two antennas are combined in a specially-designed radio receiver. As the Earth's rotation carries a radio source from east to west in the sky, the signals received at the two antennas interact with one another, sometimes adding together and sometimes subtracting. By carefully observing these fluctuations, the position of the source can be determined with much greater accuracy than can be obtained with a single antenna. And by making measurements with antennas at two or more different spacings some information about the shape and size of the source can also be obtained. Interferometers work well if the objects being observed are compact and well-separated, but may give ambiguous results for objects that are too broad or too close together. Fortunately for the radio astronomers the sky turned out to contain many distinct radio sources, so interferometry soon started to produce major results.

By 1949 Australian radio astronomers had managed to measure the positions of a number of compact radio sources well enough for them to be identified with objects seen by visible light. They were the Crab Nebula, which is the debris from a supernova explosion observed in the year 1054 AD, (see section 2.4) and two relatively nearby bright galaxies—Messier 87 and NGC 5128. A little later they identified two more radio sources—one with the nucleus of the Milky Way Galaxy, and the other with the Orion Nebula. The latter is a classic example of what astronomers call an HII region (pronounced "H-two region"), namely a cloud of interstellar ionized gas that is heated to a temperature of around 10,000 K by one or more newly-formed massive stars.

Not withstanding these successes, it was clear that the majority of the compact sources being found by interferometer surveys, including the two brightest objects in the sky, did not coincide with any bright stars or galaxies. A breakthrough came in 1951 when Graham Smith used an interferometer in Cambridge, UK, to measure their positions with an accuracy of 1 minute of arc, and persuaded Rudolph Minkowski to use the Palomar 200-inch telescope to search for visible counterparts. The brightest object, Cassiopeia A, was found to coincide with some faint filaments

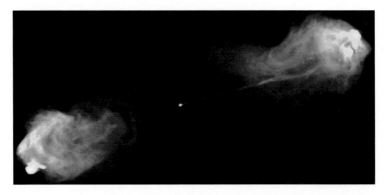

Fig. 5.7 Map of Cygnus A as made at the Very Large Array in 1984. The optically visible galaxy coincides with the small spot in the middle of the image. The outer tips of the radio lobes are about 2 minutes of arc apart, or a distance of 400,000 light years. Image credit: NRAO/AUI

that indicated it was a supernova remnant while the other source, Cygnus A, was identified with an unusually-shaped distant galaxy that was too faint to have been included in any existing catalog of galaxies. When its redshift was measured it was found to be at a distance of about 600 million light years—ten thousand times the diameter of the Milky Way galaxy. Soon other sources were also identified with even more distant galaxies and the phrase "radio galaxy" became a common term for these highly unusual objects.

Improvements in interferometry led to another important discovery about radio galaxies: in many cases the radio emission did not emanate from the galaxy itself, but from a pair of lobes on either side of it that were being pushed outward by jets of high energy particles from the galaxy, as seen in this 1984 radio image of Cygnus A (Figure 5.7). When spectroscopy was combined with photography it soon became clear that these jets were the result of some unusual energetic activity in the galaxy's nucleus. In some extreme cases these so-called "active galaxy nuclei" (AGNs) were so bright that they completely outshone the stars in the surrounding galaxy; such objects came to be known as "quasars" or "quasi-stellar objects". However, the physical processes that could produce the vast amounts of power emitted by AGNs remained mysterious until the development of X-ray astronomy and the discovery of back holes (see Chapter 9).

5.3 The Big Bang Controversy

When a new part of the electromagnetic spectrum is opened up, the initial goal is usually to search for new kinds of astrophysical objects. The discovery of radio galaxies and quasars described in the previous section is a good example of this process. As the new technology matures and observations become more reliable the time becomes ripe for careful surveys of the sky that can be used for statistical analysis.

   The most famous early radio astronomy survey is the Third Cambridge Catalog of Radio Sources, known colloquially as the 3C survey. It grew out of an earlier 2C catalog published in 1955 and emerged in its final 3C(Revised) form in 1962. It was produced by Martin Ryle and his colleagues at Cambridge University. More reliable than the earlier Cambridge catalogs it superseded, the 3CR survey contained 328 sources at 1.7-m wavelength. About 10 % of them could be identified with supernova remnants or HII regions in the plane of our Galaxy, but the majority were indisputably extragalactic. The distances to most of the sources was unknown, but several had been identified with objects at substantial redshifts, indicating that they lay at large distances from our own Galaxy.

   Martin Ryle and his colleagues argued that the 2C and 3CR catalogs could be used for a statistical test of cosmology, even if the distances to most of the sources were still undetermined. The idea was that faint sources are statistically likely to be at greater distances from us than bright ones. Because of the finite speed of light we are therefore seeing them farther back in time. So by comparing the numbers of faint and bright objects we should be able to compare the average density of the universe in the past with its average density at the present. When Ryle tested this idea he found that there were indeed relatively more faint sources than expected for a static universe, implying that the universe was expanding and getting less dense as it got older.

   Not everybody accepted Ryle's conclusion. Fred Hoyle, another astronomer in Cambridge, argued in favor of his "steady-state" theory in which new matter is continuously created to fill the gaps left by the separating galaxies. The steady state theory appealed to Hoyle because it is mathematically the simplest kind of cosmology; he criticized Ryle's data and dismissively labelled his ideas as the "Big Bang" theory, a term that has stuck, though its pejorative implications have long disappeared. The controversy remained active for many years, fading away only when it became clear that the Cosmic Microwave Background radiation, which we will discuss in Chapter 7, provided additional strong support for the Big Bang theory.

5.4 Recent Radio Surveys

Most of the major radio surveys that astronomers depend on nowadays were made using the technique of Earth Rotation Aperture Synthesis. This technique was first demonstrated in Cambridge in the 1960s, and is an extension of the principles used in interferometry. In an aperture synthesis telescope a number of modest-sized antennas are spread out over a large ground area. As the rotating Earth carries the antennas around each other the radio signals from the receivers are combined in a computer. Under the right conditions, an image of the patch of sky under study can be computed that has a resolution equivalent to that of a single dish with a

Fig. 5.8 The Very Large Array (VLA) in New Mexico. Each dish antenna is 25 m in diameter. The separation between the individual dishes can be changed to generate different desired beamsizes. Image credit: NRAO/AUI

diameter equal to the largest separation of the antennas (see Appendix A.4). This principle was used in early surveys with the One-Mile and later the Five-Kilometer Telescopes at Lord's Bridge Observatory in Cambridge. The most extensive radio sky surveys to date, however, have been produced by the Very Large Array (VLA) of the National Radio Astronomy Observatory in New Mexico (Figure 5.8).

The largest radio survey to date, the NRAO VLA Sky Survey (NVSS), was conducted between 1993 and 1998 (see Figure 5.1 and Table 5.1) with the array in its most compact configuration. It covered the whole sky north of declination $-40°$ The angular resolution of the survey (45 arcseconds) was poor by the standards of optical

Survey	Telescope	Wavelength	Beamsize	Number of sources
8C	Cambridge	7.8 m	4.5 arcmin	$\sim 6,000$
VLSS	VLA	4.0 m	80 arcsec	$\sim 70,000$
WENSS	Westerbork	92 m	50 arcsec	$\sim 200,000$
SUMMS	Molonglo	36 cm	45 arcsec	$\sim 120,000$
NVSS	VLA	21 cm	45 arcsec	$\sim 1,800,000$
FIRST	VLA	21 cm	5 arcsec	$\sim 1,000,000$
GB6	Greenbank	6 cm	3.5 arcmin	$\sim 75,000$
PMN	Parkes	6 cm	4.2 arcmin	$\sim 65,000$

Table 5.1 Major radio astronomy surveys organized by decreasing wavelength

astronomy but enabled the survey to be completed in a reasonable time and with good sensitivity. The sky south of declination $-30°$, which cannot be mapped from North America, was subsequently surveyed from Australia as the Sidney University Molonglo Sky Survey (SUMSS) at a slightly longer wavelength.

The NVSS was followed up at the VLA by the "Faint Images of the Radio Sky at Twenty-Centimeters" survey (FIRST). To perform this survey the dishes of the VLA were moved father apart, producing a narrower beam than NVSS's. Since the scientific focus of the FIRST survey is the study of radio galaxies and quasars, the survey was confined to areas of the sky well away from the galactic plane covering about a quarter of the whole sky—roughly the same area as the Sloan Digital Sky Survey (section 12.1). Observations started in 1993 and were completed in 2011.

Galaxies and quasars differ in their radio colors—in the sense of the relative strength of their emission at different wavelengths. It is therefore important to make sky surveys at a range of different wavelengths. Major long-wavelength surveys include the Netherlands-based Westerbork Northern Sky Survey (WENSS) at 92 cm, the VLA Low-frequency Sky Survey (VLSS) at 4.0 meters, and the Cambridge 8C survey which was carried out a wavelength of 7.8 meters. Several more long-wavelength surveys are in the planning or early observational stages. In the northern hemisphere we have the Low Frequency Array (LOFAR)—a network of radio telescopes based in the Netherlands and spreading over much of Europe. Long wavelength studies of the southern skies will be the responsibility of the Square Kilometer Array (SKA) under construction in South Africa and Australia.

At wavelengths shortward of 21 cm, aperture synthesis arrays such as the VLA and Westerbork become impractical for large scale surveys because of the required observing time. One approach is to make detailed maps of small areas of sky that are likely to be statistically representative of the universe in general. This approach has been taken by radio astronomers in Cambridge who have made aperture synthesis maps of several tens of square degrees of sky at 6-cm and 2-cm wavelength. The other approach to short-wavelength radio surveying is to use single dish antennas and make methodical scans of the sky. The largest such survey in the north was the GB6 survey using the NRAO's 100-m diameter antenna at Greenbank, West Virginia (Figure 5.18). The equivalent effort in the south was the Australian-based PMN survey, which used the Parkes dish (Figure 5.6).

5.5 The Pulsar Discovery Survey

In 1967 two astronomers at the University of Cambridge made one of the most unexpected discoveries in the history of astronomy. Antony Hewish, assisted by his graduate student Jocelyn Bell, had built a new kind of radio telescope that was designed to find very compact radio sources like those found in some quasars. The principle behind this telescope was to look for the radio equivalent of twinkling. If one watches a star with the naked eye its brightness often appears to fluctuate randomly, particularly if the star is close to the horizon. The twinkling is caused

Fig. 5.9 Part of the 4-acre radio telescope at Lord's Bridge Observatory with which pulsars were discovered in 1967. Image credit: University of Cambridge

by the Earth's atmosphere as the light travels through it. Thermal convection and meteorological effects cause the density, and hence the refractive index, of the air to constantly fluctuate, sometimes bending light towards the observer and sometimes away. Planets, however, do not twinkle: planets cover an extended patch of sky (typically a few tens of seconds of arc across) so their light gets to us by a variety of different paths, causing the fluctuations to be averaged out.

Radio waves are affected in a somewhat similar way when they pass through the solar system. The solar wind, which consists mainly of fast-moving clouds of ionized hydrogen atoms ejected from the Sun, refracts radio waves that pass through it. And, just like the twinkling of stars versus planets, compact radio sources such as quasars fluctuate in brightness more than extended radio galaxies such as the double lobes of Cygnus A (Figure 5.7). Hewish's plan was to measure the strength of this scintillation and use it as a quick way to establish which radio sources were quasars and which were extended radio galaxies.

The radio telescope that Hewish designed consisted of a 4-acre field of dipole antennas made of copper wire suspended on wooden posts (Figure 5.9). It monitored radio sources as they transited from east to west each day. The signals from the dipoles were combined together and recorded on strip charts in a nearby laboratory. A crucial feature of Hewish's receiver system was the speed at which data was recorded. Normally, radio astronomers try to minimize random fluctuations in their data by electronically smoothing it before recording it. This process improves what scientists call the signal-to-noise ratio; in the pre-computer era it also saved money by minimizing the amount of expensive strip-chart paper being used. Because Hewish and Bell were looking for fluctuations on a time scale of tenths of a second they ran their strip charts at high speed: one day's observation could produce 100 feet of paper that had to be laboriously examined by hand.

While Bell was searching the strip-charts for scintillating quasars, she noticed four objects that seemed to be fluctuating in a quite different way. When examined in detail their emission was found to take the form of regular narrow radio pulses between 0.25 and 1.3 seconds apart. No other object in the sky had ever been seen to behave like this, and astronomers rushed to try to come up with ideas as to what could be causing the pulses. One thing was immediately apparent: these pulsars, as the objects were soon named, had to be less than about 0.05 light-seconds across, otherwise the radio emission from different parts of the pulsating object would arrive at Earth at slightly different times, blurring the narrow pulses. Pulsars therefore had to be Earth-sized or smaller. White dwarf stars, which have diameters of about this size, were considered, but an alternative explanation, namely that a pulsar was a rapidly rotating neutron star formed during a supernova explosion, was soon widely accepted. A crucial confirmation of the supernova hypothesis was the discovery of a rapidly rotating pulsar in the centre of the Crab Nebula.

Since 1967, astronomers have discovered more than 2000 pulsars with periods ranging from 1.5 milliseconds to 8.5 seconds. Most of these were found during surveys by the Parkes radio telescope in Australia, but some pulsars have also been detected at visible, X-ray, and gamma-ray wavelengths.

While pulsars have taught us a great deal about the physics of supernova explosions, the extreme regularity of their pulses has also allowed physicists to perform precise tests of general relativity by watching pulsars that are members of binary star systems or planetary systems. The story of their discovery is also a fine lesson to scientists to be on the lookout for the unexpected, as well as anticipated phenomena in their experiments

5.6 The 21-cm Spectral Line

The existence of interstellar gas in the Milky Way galaxy was known to astronomers ever since the discovery of gaseous nebulae (see section 3.3). The gas in these nebulae is kept hot and ionized by the ultraviolet radiation from the stars they surround. But interstellar gas that is not close to a hot star is harder to detect. Since 1904 astronomers had known about faint narrow absorption lines of sodium and calcium in the spectra of distant stars: these lines provided indirect evidence that the space between stars is filled with cool gas. But direct evidence for the existence of interstellar hydrogen—a far more abundant element than calcium or sodium—was harder to obtain since cool hydrogen gas displays nether emission nor absorption lines at visible wavelengths.

In 1944 Hendrik van de Hulst, a Dutch astronomer, took a closer look at the physics of the neutral hydrogen atom and the fact that its lowest energy state is split into two levels depending on how the spins of the electron and proton are lined up. The energy difference is very small and corresponds to a wavelength of 21 cm. He estimated that there was enough hydrogen in the Milky Way that a radio telescope should be able to detect the line if it was fitted with a suitable receiver.

Fig. 5.10 21-cm signals from clouds moving at different speeds show up separately when their wavelengths, as well as their strengths, are measured

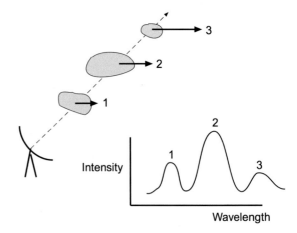

The line was first found in 1951 by Edward Purcell and Harold Ewen, using a specially built horn antenna poking out of the window of a laboratory on the Harvard campus. Their detection was rapidly confirmed by astronomers in the Netherlands and in Australia. It was quickly established that the gas is heavily concentrated along the plane of the Milky Way. It has an average density of about one atom per cubic centimeter, but there are many clouds with densities hundreds of times larger than this.

The 21-cm line can tell us about far more than just interstellar gas densities. The 21-cm photon that a hydrogen atom emits always has the same sharply defined wavelength. But if we are moving away from (or towards) that atom at the time that we detect that photon we will measure it to have a slightly longer (or shorter) wavelength. This phenomenon is called the Doppler effect; the same principle is used by police to detect speeding motorists. Figure 5.10 shows how we can use this idea in radioastronomy: the radio telescope on the left is collecting radiation from one particular direction in which there are three clouds of gas moving at different speeds. If the radio receiver attached to the telescope is appropriately designed it can measure both the strength and the wavelength of the photons it receives. In the case illustrated the three clouds will show up at slightly different wavelengths. Note that we cannot use the Doppler when observing synchrotron radiation—the cause of most of the emission seen in Figure 5.5—because we have no way of knowing the original wavelength of a synchrotron photon.

Figure 5.11 shows a 1958 attempt by Jan Oort, Frank Kerr, and Gart Westerhout to map the neutral hydrogen in our Galaxy using observations of the 21-cm line data obtained from telescopes in both the northern and southern hemispheres. To generate this map it was necessary to make some assumptions about the rotation speed of the Galaxy at different distances from the Sun. The map should therefore be regarded as indicative rather than definitive. It shows some signs of spiral arms, but a clearer untangling of the structure of our Galaxy had to wait until data on molecular clouds became available (see section 5.7).

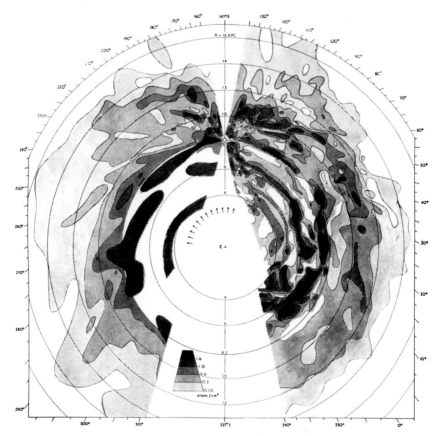

Fig. 5.11 The distribution of neutral hydrogen in the Milky Way galaxy as determined from 21-cm line observations in Netherlands and Australia in 1958. The V-shaped areas centered above and below the location of the Sun are left blank because the Doppler shifts of the line are too small to use. Image ©Royal Astronomical Society/Oxford University Press

Although the map is somewhat disappointing in its detail and in its ambiguity it should be remembered that observations of radio spectral lines, together with infrared mapping, provide us with almost all the information we possess about many parts of the Milky Way galaxy. Because of interstellar dust extinction it is difficult to study the light from stars more than a few hundred parsecs away from us in the galactic plane—a small fraction of the estimated 8 kpc to the center of the Galaxy.

We have clearer pictures of the neutral hydrogen in some external galaxies than we do for our own. Figure 5.12 shows a map of the neutral hydrogen in the spiral galaxy M81 made using the VLA; the spiral arms are much more clearly laid out in this galaxy than in our own.

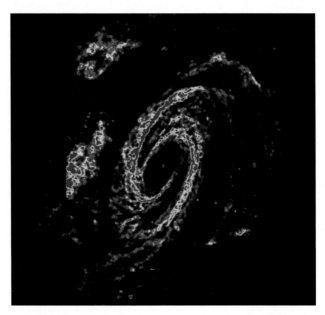

Fig. 5.12 21-cm line emission from the spiral galaxy M81. See Figure 8.3 for an image of the same galaxy at visible and ultraviolet wavelengths. Image credit: NRAO/AUI

5.7 Molecules in the Milky Way

As we noted in the previous section, hydrogen in its atomic form is widespread in interstellar space. But when the density of hydrogen atoms gets large enough they tend to combine to form H_2 molecules, which produce neither the 21-cm line, nor any other spectral lines at wavelengths that can penetrate dense, dusty regions of space. In the 1960s, therefore, astronomers who wanted to study how new stars form out of interstellar gas started looking for other molecules that might be mixed with the H_2, and which had spectral lines they could observe with radio telescopes. Four radio spectral lines of the OH molecule were discovered at 18 cm wavelength in 1963, but the physics of these transitions proved to be so complicated that little could be learned from them. Much more progress was made in the 1970s–80s as radio astronomers pushed their telescopes and receivers to shorter and shorter wavelengths where many more molecules have spectral lines.

A major step forward came in 1970 with the discovery of radio emission from the carbon monoxide molecule (CO) by Robert Wilson, Keith Jefferts, and Arno Penzias at Bell Laboratories. They observed a strong spectrum line from the molecule at a wavelength of 2.6 mm when they pointed their telescope first at the Orion Nebula, and then in the direction of several other known concentrations of interstellar gas. Considerations of the chemistry of interstellar gas indicated that there should be a fairly constant ratio of about 10^4 hydrogen molecules for every CO molecule in a

Fig. 5.13 Dark nebulae such as Barnard 68 in the Ophiuchus constellation contain molecular hydrogen and dust grains which hide our view of the stars behind it. Image credit: European Southern Observatory

dense molecular cloud. Maps of the 2.6-mm CO emission are therefore a good proxy for maps of the distribution of the far more abundant molecular hydrogen gas.

Follow-up studies led to the realization that large clouds of molecular gas are common in the Milky Way galaxy. While it is their carbon monoxide molecules that make them easy to observe, 99 % of their mass is comprised of hydrogen molecules and helium atoms. Dust grains mixed with the gas make these clouds opaque to visible light, causing some small ones to appear to us as dark nebulae (Figure 5.13).

The absence of starlight energy in these dark clouds causes their temperatures to fall to around 10 K. The resulting drop in gas pressure in their centers can make them unstable against their own internal gravitational forces. The cloud may then collapse under its own weight, forming one or more new stars. Figure 5.14 shows the result of mapping the CO emission in the Orion constellation: the vast molecular clouds indicated by the red contours were unknown before 1970, but help us understand why there are so many young stars clustered in this region. The Orion Nebula itself (Figure 4.1) is now seen as a place where a group of four recently formed stars (the Trapezium cluster) is eating away at the molecular cloud, sending newly-ionized gas rushing towards us as it heats up the dense cold gas behind it.

The most extensive survey for molecular clouds was carried out over a period of nearly 20 years ending in the year 2000 by Thomas Dame, Dap Hartmann, and Patrick Thaddeus. They employed two radio telescopes, one at Harvard College in the northern hemisphere, and one in Chile in the south. Even though each antenna was only 1.2 meters in diameter, the short wavelength of the CO spectral line gave them sufficient angular resolution for their task. Their final map, in Figure 5.15,

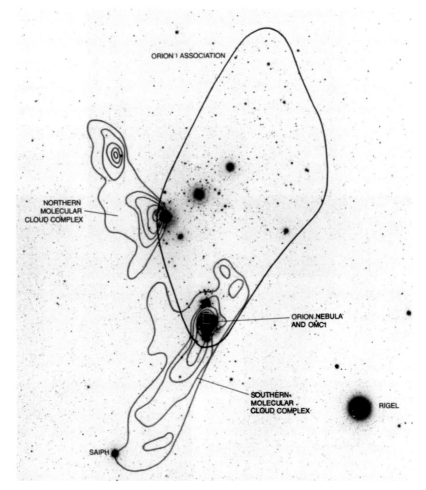

Fig. 5.14 Negative image of the southern part of the Orion constellation, including Orion's belt and the Orion nebula. The black contour shows the extent of the Orion I association of young stars while the red contours show the molecular clouds as revealed by observations of the carbon monoxide molecule

shows that the molecular clouds are very strongly concentrated to the narrow plane of the Galaxy, and that the vast majority of the clouds are closer to the center of the Galaxy than we are. The lower image in Figure 5.15 shows the very similar 100-μm emission from warm dust in the Galaxy as measured by the IRAS infrared satellite (see section 6.4), indicating that the gas in these clouds is mixed with warm interstellar dust particles. The emission from the 21-cm atomic hydrogen line, which is a marker of low-density interstellar gas, is much more widespread than that from the CO line.

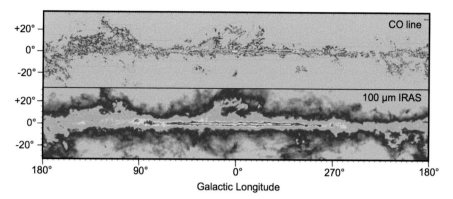

Fig. 5.15 Map of the 2.6-mm carbon monoxide emission from the plane of the Milky Way galaxy compared with a map of the 100-micron hot dust emission obtained by the IRAS infrared survey. Image ©American Astronomical Society, modified by author

5.8 Searching for ET

The can be few astronomers who have never asked themselves the question "Is there life elsewhere in the universe?" There are two astronomical approaches to answering this question. The first is the science of astrobiology, which involves learning as much as possible about the earliest lifeforms on Earth, and devising ways to look for related chemical signatures in the solar system and in nearby planetary systems. Astrobiologists' searches are of necessity confined to a small region of space within and near to the solar system, but their research enthusiasm would be vindicated by the discovery of a life form as simple as slime.

The second approach to finding extraterrestrial life in the universe is to search for celestial signals that show patterns of non-randomness that cannot be explained by any natural process. Such a signal could only be produced by something operated by or constructed by some kind of intelligent being. Let us follow Hollywood and refer to this hypothetical form of intelligent life as "ET" and ask ourselves what kind of signals we should look for. The laws of physics tell us that the most efficient and fastest way of sending signals through space is by using electromagnetic waves, which then leads to the question: what wavelength of electromagnetic waves would ET use? There are three arguments that lead us to conclude that we should look for ET with radio telescopes:

Fig. 5.16 The 305-m diameter radio telescope at Arecibo, Puerto Rico. Image credit: Wikipedia

- The interstellar medium is transparent to all radio wavelengths except the very longest. It becomes progressively more opaque as we move to the shorter visible and ultraviolet wavelengths. It becomes transparent again to X-rays and gamma rays.
- ET most probably lives in a planetary system where he can make use of the energy from a star. The infrared, visible, and ultraviolet light from his star will most likely dazzle any signals that he may be generating at these wavelengths.
- The laws of quantum physics tell us that far more energy is required to produce a photon of X-ray or gamma-ray radiation than a photon of radio radiation (see Appendix A.3). It is therefore more cost-effective to send radio signals than, say, X-ray signals.

The first person to make a serious search for radio emission from ET was Frank Drake who, in 1960, used the 25-m diameter antenna of the National Radio Astronomy Observatory at Greenbank, Virginia to search for signals from two nearby stars, Epsilon Eridani and Tau Ceti. He did not detect anything of interest, nor have a number of other searches made with other radio telescopes in the subsequent decades.

Several strategies have been used to search for ET. One approach is simply to look at as many reasonably nearby stars that one can get the telescope time for. This was the approach taken by Project Phoenix between 1996 and 2004. The project,

Fig. 5.17 The Allen Telescope Array in northern California. Image credit: Adam Hart-Davis

which made use of radio telescopes in Australia and the USA and was entirely privately funded, observed about 800 stars, all within a distance of 200 light years.

A second approach is to let the telescope point at a random direction in the sky and then analyze the resulting signal for any interesting patterns. This approach is being used by the SERENDIP project on the Arecibo radio telescope in Puerto Rico (Figure 5.16). This telescope consists of a 305-m diameter dish carved out of a valley. The direction the telescope is pointing is set by adjusting the position of a radio receiver that is suspended high above the dish. For most radio astronomy observations the receiver is carefully tracked sideways above the dish to keep the focus of the telescope pointing at a particular object as it transits from east to west. The SERENDIP project uses a second receiver on the telescope that points in a different and randomly changing direction in the sky while the telescope is performing its main scientific program. This "piggy-back" approach allows the project to get its observing time for free. Signals from this second receiver are then collected and distributed for analysis to thousands of personal computers via the SETI@home crowd-sourcing scheme.

A more recent contribution to the search for ET is the Allen Telescope Array, in northern California. It consists of 44 6-m diameter dishes that can be pointed anywhere in the sky. It is operated by the SETI Institute. This instrument collects signals at a very wide range of wavelengths and, if sufficient financial support can be found, may eventually be expanded to 350 dishes. (Figure 5.17). The instrument also carries out a wide range of conventional radio astronomy programs.

Fig. 5.18 The 100-m diameter NRAO Green Bank Telescope. Image credit: NRAO/AUI/NSF

In late 2015 it was announced that the Russian entrepreneur Yuri Milner has donated $100 million to fund a 10-year search for intelligent life. The first initiative is called "Breakthrough Listen" and will use two very large single-dish radio telescopes, namely the 64-m diameter Parkes radio telescope in Australia (Figure 5.6) and the 100-m diameter fully steerable dish at the US National Radio Astronomy Observatory at Greenbank West Virginia (Figure 5.18). Over a period of ten years the telescopes will be used to search for faint signals from a million of the closest stars to Earth. It will also make scans of the center of the Milky Way galaxy and 100 nearby galaxies.

Some scientists believe that the search for ET is unlikely to be a success. This idea was first expressed in 1950 by the physicist Enrico Fermi who argued that if ETs exist, some of them should be so much more advanced than us that they would have developed the technology to explore and colonize planetary systems all over the Galaxy. The fact that we see no obvious signs for their existence should therefore be interpreted as evidence that they do not exist. This argument is known as the Fermi Paradox.

Chapter 6
Infrared Surveys

In the year 1800, William Herschel, whose studies of Uranus, double stars and nebulae we have already discussed, made what many would consider the most important discovery of his life, namely infrared radiation. It happened while he was studying the heating effect of different colors of light that had been split by a prism. He noticed that the thermometer he was using to analyze sunlight registered the strongest heating effect beyond the red end of the spectrum. His follow-up experiments showed that the heating effect was caused by radiation that could be reflected, refracted, and absorbed in a similar way to light. He called this radiation Calorific Waves; the word infrared came later.

What astronomers now call the infrared waveband extends from 0.7 µm, the long-wavelength limit of the human eye, to about 1 mm (1,000 µm), where the radio band takes over (Chapter 5). The Earth's atmosphere blocks out radiation at many of these wavelengths but between 0.7 and 35 µm and between 350 µm and 1000 µm there are a number of transparent atmospheric "windows" that allow astronomers to observe with ground-based telescopes using special filters (see Figure 6.1). Some of these windows, including those at 1.6 µm and 2.2 µm, are almost perfectly transparent even down to sea-level, but others, such as those at 20 µm, 30 µm, and 350 µm, are strongly affected by the amount of water vapor in the atmosphere. For meteorological reasons, water vapor tends to be concentrated at low altitudes, which is why specialized infrared telescopes are usually located on the summits of high mountains such as Mauna Kea in Hawaii at 4,200 m altitude. Observations between 35 µm and 350 µm have to be made from even higher altitudes using special telescopes carried by aircraft, balloons or satellites. Astronomers customarily refer to the range from 0.7 µm to 2.5 µm as the "near-infrared," the range from 2.5 µm to 35 µm as the "mid-infrared," the range from 35 µm to 350 µm as the "far-infrared," and the range from 350 µm to 1,000 µm as the "submillimeter" though those are not strict definitions.

© Springer International Publishing Switzerland 2016
G. Wynn-Williams, *Surveying the Skies*, Astronomers' Universe,
DOI 10.1007/978-3-319-28510-8_6

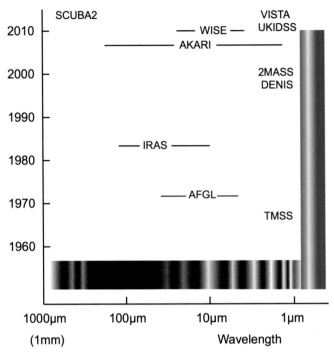

Fig. 6.1 Major infrared surveys sorted by approximate dates and wavelength ranges. The shaded bar near the bottom of the figure shows how the transparency of the atmosphere changes with wavelength: black means the atmosphere is opaque: white or gray indicates at least some transparency

6.1 The Caltech TMSS Survey

Detecting faint infrared signals is not easy, and early astronomical progress was slow. Charles Piazzi-Smyth detected the Moon in 1856 but the first reliable detection of stars—Vega and Arcturus by Ernest Nichols—had to wait until the beginning of the 20th century. Methodical studies of the infrared properties of stars began in the early 1960s, triggered by the development of sensitive solid-state infrared detectors, such as lead sulfide photoconductors and germanium bolometers. Early infrared studies, notably by Harold Johnson of University of Arizona, were focused on understanding the radiation from the photospheres of cool stars and understanding the dimming of light by interstellar dust. Infrared astronomy at this time was generally regarded as a straightforward extension of visible light photometry.

The real blossoming of infrared astronomy, like that of radioastronomy, came not from astronomers themselves but from physicists. At the California Institute of Technology, Gerry Neugebauer and Robert Leighton decided to apply their skills to making a survey of the whole sky at 2.2 μm. This later became known as the TMSS, for Two Micron Sky Survey.

Knowing that they would never get enough observing time on any existing large telescope, they designed and built their own instrument from scratch. Since the main role of their telescope mirror was to collect plenty of photons rather than produce perfect images, they made their mirror out of lightweight epoxy resin rather than glass. To get the required parabolic surface they poured liquid epoxy into a 62-inch diameter mould, which they then spun horizontally at a steady speed with an electric motor. The centrifugal force resulting from the rotation caused the epoxy at the outer parts of the mould to rise slightly as it set, giving the whole surface the shape of a paraboloid. After the epoxy had set, it was coated with a thin layer of aluminum, making a concave mirror with a focal length of about 60 inches. The mirror was installed on a specially-designed telescope, (see Figure 6.2) which focussed the infrared radiation onto a group of eight lead sulfide detectors that were cooled by liquid nitrogen. It performed its sky survey from Mount Wilson Observatory in California.

A problem that plagues all infrared astronomy is that detectors inevitably pick up radiation from the Earth's atmosphere and from the telescope system itself. Neugebauer and Leighton eliminated most of this background radiation by having their whole telescope oscillate 20 times a second between two patches of sky

Fig. 6.2 Neugebauer and Leighton's 2.2 μm telescope. Adapted from image NASM 98-15667 from the Smithsonian National Air and Space Museum

5 arcminutes apart and recording only the differences in the brightness between the two patches of sky. This technique, known as chopping, works because the emission from the Earth's atmosphere is reasonably uniform on small angular scales. The signals were recorded on a strip chart recorder; computers at that time were still far too bulky and expensive to be installed at an observatory.

Observations were carried out and analyzed, largely by Caltech students, over a three-year period starting in 1965, resulting in a catalog of 5562 stars brighter than 3rd magnitude in the declination range $-33°$ to $+81°$. Sources in it have the designation IRC for Infra Red Catalog.

As expected, the majority of objects in the TMSS turned out to be stars that had already been catalogued at visible wavelengths. However, the ratio of 2.2-μm brightness to visible brightness covered a wide range; stars with cool photospheres, such as those with spectral types M or K, emit the bulk of their power in the infrared; stars with hot photospheres such as spectral types O or B are relatively weaker in the infrared. This difference is a natural consequence of Wien's law in physics, that the peak wavelength at which an object radiates thermally is inversely proportional to temperature (see Appendix A.9).

The most extreme example of an infrared-bright star was IRC+10216, which is fainter than 18th magnitude in the visible, but at 5 μm is the brightest object in the sky outside of the solar system. Follow-up observations showed that it has a diameter of a few seconds of arc and varies on a timescale of about 600 days. The infrared spectrum of the object indicated that whatever was causing the emission was at a temperature of about 650 K—far cooler than the surface of any star previously observed.

What could cause the infrared emission from objects such as IRC+10216? The explanation that has emerged is that these objects are old stars that are approaching the ends of their energy-generating lifetimes. Such stars go through a stage of rapid mass loss, ejecting large amounts of matter into space in the form of stellar winds. The hot gas in the winds cools as it moves away from the star and some atoms, among them carbon, silicon, and oxygen, combine and condense to form small grains of dust that surround the star like a shell. The dust grains then absorb part of the light from the star, becoming gently heated to temperatures in the range 100–1000 K and producing the infrared emission. In some cases, such as IRC+10216, this dust shell is so thick that the star itself becomes almost completely hidden at visible wavelengths; in others the star and the dust shell both shine, albeit mainly at different wavelengths. This process of mass loss by cool stars is now recognized as the main origin of the dust particles that are the cause of interstellar extinction and reddening (Appendix A.8). It also explains the origin of planetary nebulae and is the main mechanism by which matter is recycled from old stars to make new ones. Some idea of the importance of the TMSS in stimulating the study of mass loss from cool stars may be gauged from the fact that more than 900 scientific papers have since been published about IRC+10216 alone.

Neugebauer and Leighton's hunch that "you will find something new and interesting by looking at the sky at a new wavelength" was thoroughly vindicated.

6.2 2MASS and DENIS

As we shall discuss in the next two sections, the main thrust of infrared surveys in the 1970s and 1980s was in the direction of longer wavelengths and space-based exploration. But by the mid 1990s it was clear that detector technology had progressed so much that it was time for a far more sensitive ground-based infrared survey. In fact, two were carried out almost simultaneously; the 2MASS survey and the DENIS survey.

2MASS stands for Two Micron All Sky Survey, a particularly appropriate acronym since two of its lead scientists, Susan Kleinmann and Michael Strutskie, were based at UMASS—the University of Massachusetts. Some of the parameters of this survey are shown in Table 6.1, which also includes data for the earlier Caltech 2.2 μm TMSS survey.

The 2MASS and DENIS surveys were broadly similar in their scope, but each had some complementary advantages: 2MASS had better sensitivity and better sky coverage, while the DENIS survey covered a wider range of wavelengths. Both surveys reflect the enormous technical advances in infrared technology that arose during the 30 years since 1968. These improvement came in two main areas: infrared detectors and computing power.

Early infrared astronomy observations were made using either a single detector at the focal plane of the telescope or a handful of detectors at most. Infrared-sensitive semiconductor arrays, first developed for the US military, started to become available to astronomers in the late 1980s. They grew rapidly in both size and sensitivity so that by the time the 2MASS and DENIS instruments were being built, solid-state arrays of 256×256 pixels were available. These semiconductor detectors were made out of a subtle combinations of the metals mercury, cadmium, and tellurium cooled to the temperature of liquid nitrogen. Each of the 65,000 detectors was far more sensitive than each of the 8 detectors in the Caltech 2.2 μm survey, so the overall sensitivity improvements were enormous: the faintest source in the 2MASS survey is 30,000 times fainter (11.3 magnitudes) than the faintest source in the Caltech survey. This gain comes despite the fact that both surveys used telescopes of roughly the same size.

Survey name	TMSS	2MASS	DENIS
Observing Epoch	1965–1968	1997–2001	1996–2001
Observing site	California	Arizona and Chile	Chile
Coverage	$-33°$ to $+81°$	All-sky	$-88°$ to $+2°$
Wavelengths (μm)	2.2	1.25, 1.65, 2.2	0.85, 1.65, 2.2
Telescope Size (m)	1.6	1.3	1.0
Pixel size (arcseconds)	600×180	2	3
Number of 2.2-μm detectors	8	65,536	65,536
Limiting 2.2-μm magnitude	3.0	14.3	14.0
Number of sources	5,600	470,000,000	355,000,000

Table 6.1 Comparison of the TMSS, 2MASS, and DENIS surveys

Neither the DENIS nor the 2MASS survey could have taken place before the advent of cheap computing. The telescopes were controlled by computers and the data was collected on computers in the telescope domes and stored on magnetic tapes which, in the case of 2MASS, were mailed to a data-reduction center in California.

The small sizes of the individual detector elements in 2MASS and DENIS make it possible to distinguish the infrared emission from stars (which appear point-like) and broader objects, whose images are spread over several pixel elements. While the bulk of the objects detected in these surveys are starlike, each survey yielded over a million extended objects, most of which are external galaxies.

The upper image in Figure 6.3 shows the distribution of star-like objects in the 2MASS survey. This image dramatically demonstrates one of the major advantages of an infrared survey, namely its ability to penetrate dust clouds in our Galaxy. The sky at 2.2-μm is a far better representation of the arrangement of the stars in the Milky Way than is the view of the sky we get using visible light (lower image). What Figure 6.3 shows us immediately is that the great majority of stars in our Galaxy are confined to a narrow plane in the Milky Way, and that the Milky Way itself is concentrated in one patch of the plane that surrounds the nucleus of our Galaxy. The concentration of stars in this direction (which corresponds to the constellation Sagittarius) was unknown before the advent of infrared astronomy. Indeed, before the advent of radio astronomy, estimates of the direction of the Galactic Center had been wrong about 30°. Astronomers who study stars in the Milky Way at visible wavelengths are limited by interstellar extinction to those within a few hundred parsecs of the Sun; at 2.2 μm they can see ten times farther before dust blocks their view (Appendix A.8).

Figure 6.4 shows a map of the galaxies found in the 2MASS survey. The map was made by eliminating all the sources that were point-like and therefore probably stars. The individual galaxies in this image are not scattered randomly but are concentrated into clusters and streams on many different size scales. Since the 2.2-mum radiation from a galaxy is a better guide to its mass than is its visible light, images such as this, when combined with redshift estimates of a galaxy's distance, provide the best data we possess for the distribution of matter in our local universe.

Another important advantage of the 2MASS and DENIS surveys is their ability to detect low-mass stars in the Milky Way galaxy. Low mass stars tend to have low surface temperatures; they therefore radiate at longer wavelengths than solar-type stars and stand out much better at infrared wavelengths than at visible wavelengths (see Appendix A.9). Of particular interest are the brown dwarf stars, which are objects with masses intermediate between sun-like stars and Jupiter-like planets. They are warm because they have retained much of the heat released during their formation billions of years ago, but their internal temperature and pressure is too low for thermonuclear nuclear reactions to take place. Before 2MASS and DENIS very few brown dwarf stars were known, but it is now thought that they might be as common as stars.

Fig. 6.3 All-sky map of the point sources seen by 2MASS (upper figure) compared with a similar view at visible wavelengths (lower figure). These images are in Galactic Coordinates, so are centered on the nucleus of Milky Way galaxy. The colors blue, green, and red in the upper image correspond to 1.25, 1.65, and 2.2 μm. Image credit: 2MASS/UMass/IPAC-Caltech/NASA/NSF (upper panel): Lund Observatory, Sweden (lower panel)

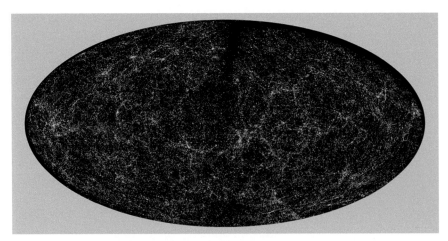

Fig. 6.4 2MASS map of the galaxies in the local part of the universe. Image credit: 2MASS/UMass/IPAC-Caltech/NASA/NSF

6.3 The AFGL survey

It is possible to study small patches of sky from the ground at wavelengths in the 3–20 μm range, but to survey the whole sky at wavelengths longward of 2.2 μm, astronomers must resort to space-borne or air-borne telescopes. There are two quite separate reasons for this.

The first problem is atmospheric transmission. Although there are some transparent bands between 2.2 μm and 35 μm (see Figure 6.1) there are none between 35 μm and 350 μm that can be used from sea level or even from mountain-top observatories. Balloon-borne telescopes and high-flying aircraft such as NASA's SOFIA observatory can be used to study interesting patches of sky at certain wavelengths in this range, but they cannot provide the necessary observing time to make an all-sky survey.

The second problem is background radiation. All solid objects and gases emit electromagnetic radiation over a wide range of wavelengths. As described in Appendix A.9, the wavelength of maximum emission depends on the object's absolute temperature: since room temperature objects on the surface of the Earth have temperatures of around 300 K, we find ourselves immersed in a sea of infrared radiation that peaks at a wavelength of 10 μm. This background radiation is the greatest problem faced by infrared astronomers.

Background radiation is emitted from the detectors and the telescope and anything else in the line of sight between the detectors and the stars; these include the mirrors and lenses that focus the light, any supporting girders that hold them in their place, and the molecules in the Earth's atmosphere. It is not difficult to cool the infrared detectors themselves: indeed some devices used by astronomers are routinely cooled to a fraction of a degree above absolute zero. But drastically cooling the primary mirror of a ground-based telescope is out of the question: its surface would be rapidly covered by an ever-fluctuating layer of condensed ice crystals, or even by liquid air, and its ability to focus radiation would be ruined. The only way to make use of a cooled mirror is to put it far into space where there is negligible atmosphere to condense on it.

The first people to make an infrared survey using a space-borne cooled telescope were Stephan Price and Russ Walker of the Air Force Cambridge Research Laboratory (AFCRL) later renamed the Air Force Geophysical Laboratory (AFGL). In the early 1970s the U.S. Air Force was developing an infrared-based ballistic missile detection system, and needed to know the locations of cosmic infrared sources that might confuse it. Price and Walker built a 16.5-cm diameter infrared telescope that was launched into a sub-orbital trajectory by an Aerobee rocket (Figure 6.5), returning to Earth a few minutes later by parachute.

Fig. 6.5 AFGL Aerobee rocket payload showing the telescope in launch position (*left*) and in an observing position (*right*), The overall diameter of the rocket is 45 cm; the telescope mirror diameter is 16.5 cm. Image credit: Springer Publishing Company

The telescope's mirror, as well as the detectors, was cooled to a few degrees above absolute zero using liquid helium. By doing this, Price and Walker were able to reduce the background radiation hitting the detectors by a factor of over 1,000,000. The resulting increase in sensitivity meant that they could detect infrared sources in a tiny fraction of the time it would take for a room-temperature telescope. The sky's radiation was focused onto an array of 24 germanium detectors which collected radiation at $11\,\mu m$ and $20\,\mu m$, plus either $4\,\mu m$ or $27\,\mu m$. The telescope was launched on nine occasions between 1971 and 1974; seven of the launches were made from New Mexico and two from Australia, providing coverage in both hemispheres. Each launch provided 200 seconds of observing time, during which time the payload was spun around a roughly vertical axis about ten times a minute with the telescope pointing at a slightly different elevation each revolution.

All of the northern sky and most of the southern sky were successfully surveyed by this method, leading to final catalog of over 3,000 infrared sources. While many of the subjects detected in the survey could be identified with bright stars in the Caltech $2.2\,\mu m$ catalog, about 30 % could not; understanding the nature of these newly discovered objects was one of the main activities of infrared astronomers in the late 1970s and early 1980s. Since they now knew where to look, astronomers could use large ground-based telescopes to map the sources with much finer detail than the rocket observations. They could also use an infrared spectrograph to look for emission or absorption lines in the spectrum of the object.

The initially unidentified objects in the AFGL catalog showed a wealth of different properties, but it was soon clear that almost all could be attributed to dust grains that were heated by nearby stars of some kind. While some of them were old stars, many others were found to lie close to the galactic plane and tally well with existing maps of radio continuum emission from HII regions and maps of 2.6-mm CO spectral line (see section 6.3). This was not a total surprise, as infrared astronomers using ground-based telescopes had already been using radioastronomy maps to guide them to HII regions, which are strong infrared sources. There they could study not only the warm dust that was mixed with the ionized gas, but other nearby infrared sources such as protostars. In both these cases the infrared emission arises from dust mixed with the gases out of which new stars have been recently forming.

The AFGL survey convincingly demonstrated the enormous sensitivity advantages that could be gained by using a cooled space-borne telescope. It therefore played a major role in building the case for a dedicated satellite for infrared astronomy—a remarkable achievement for a survey that used a total of only 30 minutes of observing time using a telescope smaller than that owned by many amateur astronomers.

6.4 IRAS

By the mid 1970s it was clear that the time was ripe for a satellite-based all-sky infrared survey, and that the survey should extend to wavelengths as long as 100 μm. The justification for the longer wavelengths was the tantalizing data that had been collected by several teams from groups in the UK and the USA, who were using balloon-borne telescopes to map selected regions of the sky. They had shown that the galactic plane, at least, contained many very powerful sources of radiation at these longer wavelengths.

In 1976 a group of astronomers from the USA, the Netherlands, and the UK got together and started working on the Infrared Astronomy Satellite, or IRAS (Figure 6.6). They were led by Gerry Neugebauer from Caltech and by Reiner van Duinen from the Netherlands. The core of the IRAS satellite was a 57-cm diameter liquid-helium-cooled mirror which focussed radiation onto a group of 62 cooled detectors, Each detector collected radiation in one of four wavelengths, 12, 25, 60, and 100 μm.

The satellite was successfully launched in January 1983 and placed into a special orbit that followed the boundary between the sunlit and dark hemispheres of the Earth. The IRAS telescope always pointed away from the Earth with the Sun at 90° to one side: in this way the satellite was subject to the same amount of heating from the Sun throughout the 24 hours of the day, while slowly scanning the whole celestial sphere. When the liquid helium finally ran out, ten months later, IRAS had mapped 96 % of the sky.

IRAS was a spectacular success with over 250,000 sources catalogued. It showed, once and for all, that the infrared sky was vastly different from the visible sky and vastly different from the radio sky. Few surveys have ever impacted such a breadth of astronomical disciplines, including comets, stars, interstellar matter, and galaxies. After IRAS, no astronomer could ignore the infrared waveband.

Fig. 6.6 Artist's impression of IRAS in orbit. Image credit: NASA/JPL-Caltech

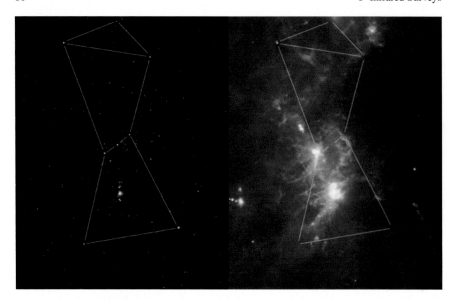

Fig. 6.7 IRAS map of the infrared emission from the region of the Orion constellation (*right*) compared with a visible light image (*left*). The bright patch in the "Sword" region of the visible constellation is the Orion Nebula (Messier 42). Image credit: NASA/JPL-Caltech/IRAS/H.McCallon

Figure 6.7 shows an IRAS map of the infrared emission from the region of the Orion constellation. The intricate patterns that fill the image are called "infrared cirrus" and arise from interstellar dust heated by ambient starlight. This is the dust that causes normal interstellar extinction. For lack of other information, astronomers had previously assumed that interstellar dust is spread out more or less uniformly, but IRAS showed that it was distributed in a complex pattern reminiscent of terrestrial cirrus clouds.

The two brightest regions in Figure 6.7 are places where new stars are being formed; they correspond to the clouds of molecular hydrogen found using radio surveys of the CO molecule (Figure 5.14). These clouds are usually so thick and dusty that even when they contain newly formed stars they emit no light that we can detect. In the brightest infrared region, however, some of the newly formed stars have pushed away the surrounding dust revealing the famous Orion Nebula. The Orion Nebula region is a place where large, bright stars are being formed, but IRAS also revealed many smaller clouds containing young and partially-formed stars that might well resemble precursors of stars like our own Sun. IRAS also found that many mature stars have a disk of dust surrounding them, lending strength to the then untested hypothesis that stars other than the Sun might have planetary systems.

Astronomers had been carefully studying the far-infrared emission from selected galaxies ever since Frank Low and George Aumann detected the Seyfert galaxy NGC1068 at 50–300 μm in 1970. But few of them expected the wonderful flood of infrared galaxies that resulted from the IRAS survey. IRAS detected roughly 25,000 galaxies. The vast majority of them are spiral galaxies and only a few are elliptical, which is not surprising considering that elliptical galaxies are known to contain very little interstellar dust. Hundreds of galaxies were discovered that emitted more than 95 % of their total luminosity in the infrared; these galaxies came to be known as starburst galaxies. Follow-up studies showed that many of these starbursts were actually colliding spiral galaxies in which rapid star formation had been triggered by the mutual impacts of molecular clouds from the two galaxies. Some of these starburst galaxies are as luminous as the most luminous quasars.

IRAS also made discoveries within the solar system. The term "zodiacal dust" refers to the dust particles that orbit the Sun in the ecliptic plane. They are the cause of the zodiacal light, which is a faint glow of scattered sunlight in the sky that can be observed with the naked eye on a dark night just after evening twilight or just before dawn twilight. IRAS detected an infrared glow from these zodiacal dust particles indicating that they had been warmed by the sunlight they had absorbed. The infrared energy radiated by the particles greatly exceeds the light energy that they scatter, implying that the zodiacal dust particles must be good absorbers of light. IRAS also found evidence that some of the zodiacal dust had been formed fairly recently—probably as a result of collisions between asteroids within the last few million years. Finally, IRAS discovered six new comets, including one, Comet IRAS-Araki-Alcock, that passed within 5 million kilometers of the Earth.

6.5 AKARI and WISE

In the years after the IRAS survey, several infrared space telescopes were launched, including the European Infrared Space Observatory (ISO) in 1995, NASA's Spitzer Space Telescope in 2003, and the European Herschel Telescope in 2009. These observatories allowed astronomers to make detailed and complex studies of chosen regions of the sky, but their fields of view and their lifetimes were much too small for them to survey the whole sky.

Two all-sky successors to IRAS were launched in the late 2000s, AKARI by the Japanese in partnership with the Europeans, and WISE by NASA. AKARI, named after the Japanese word for light, was a very versatile satellite that covered the wavelength range 1.8–180 μm and included spectroscopic instruments for studies of selected regions as well as cameras for scanning the whole sky at the longer wavelengths. AKARI was launched in February 2006 and operated at full sensitivity until August 2007, when its liquid helium ran out. It then continued to observe at a more limited range of wavelengths until November 2011.

The outstanding feature about AKARI from a survey point of view was its coverage of the far-infrared wavelength range between 65 and 160 μm, where molecular clouds, HII regions and starburst galaxies emit the bulk of their power. Despite the fact that technical considerations limited the number of far-infrared detectors to only 140, AKARI detected 400,000 infrared sources in these wavebands, and produced a beautiful map of the far-infrared sky (Figure 6.8). AKARI also surveyed the whole sky at 9 and 18 μm logging over 800,000 sources.

Fig. 6.8 Map of the sky by AKARI in galactic coordinates. Emission at 140 μm is in blue and at 90 μm is in cyan. The bright area in the galactic plane to the left of center is the Cygnus X star formation region. Infrared "cirrus" emission covers most of the sky. Image credit: Japanese Aerospace Exploration Agency and European Space Agency

The Wide-field Infrared Space Explorer (WISE) spacecraft was launched by NASA in December 2009 and mapped the sky at four wavebands between 3.4 and 25 μm during 2010. The satellite was about the same size as IRAS, but benefited enormously from the improvement in detectors over the previous 25 years. Instead of individual detectors, it was fitted with four 1048×1048 pixel infrared arrays giving it a 1,000-fold advantage in sensitivity over IRAS and a beam area that was about 1000 times smaller.

A major motivation for the WISE survey was to study asteroids. The amount of visible light we detect from an asteroid depends on its size and its albedo (the ratio of reflected to absorbed light). An asteroid whose visible light we observe might equally well be small and shiny or large and dark. But if we can measure the infrared radiation as well as the reflected light from the asteroid we can derive both the albedo and the size, since the large dark asteroid will radiate more infrared power that the small shiny one. WISE was able to measure diameters and albedos for some 100,000 asteroids—data that is currently being used to study the different populations of asteroids in the solar system.

WISE's data are also being used to study other astronomical problems such as starburst galaxies and dust around nearby stars similar to that which produces the zodiacal light in our solar system. As Figure 6.9 illustrates, WISE could also produce beautiful maps of the warm interstellar dust in star formation regions in our Galaxy.

WISE's main survey ended when the liquid hydrogen coolant ran out, and the satellite was put into a hibernation mode in 2011. But the Chelyabinsk meteorite impact in 2013 prompted renewed concerns about potentially dangerous asteroids and WISE was revived under the new name, NEO-WISE—a clever pun, since NEO

Fig. 6.9 WISE map of the infrared emission from the W3 and W4 interstellar clouds. These are regions in which clusters of hot massive stars are being born. Image credit: NASA/JPL-Caltech/UCLA

stands for Near Earth Object as well as implying rejuvenation. Although not as sensitive as it was when it was cooled by liquid hydrogen, NEO-WISE can still use its 3.4 and 4.6 μm detectors to look for asteroids it missed in its first survey if they are now making a pass relatively close to the Earth.

6.6 UKIDSS and VISTA

The most recent ground-based infrared sky surveys are UKIDSS in the northern hemisphere and VISTA in the southern hemisphere. Both operate in the 1.2–2.2 μm region, and use large solid-state detector systems. The surveys are the successors to the 2MASS and DENIS surveys discussed in section 6.2, but are also cousins to the large visible-light surveys that we will discuss in Chapter 12.

UKIDSS, which stands for UKIRT Infrared Deep Sky Survey commenced in 2005 using the 3.8-meter United Kingdom Infrared Telescope (UKIRT) on Mauna Kea, Hawaii (Figure 6.10). UKIDSS covered about 7,500 square degrees of the northern hemisphere—about 18 % of the whole sky—to a depth some 3 magnitudes fainter than 2MASS (see section 6.2). A few smaller areas were scanned with greater sensitivity. The major parts of the survey were the galactic plane as visible from

Fig. 6.10 The 3.8-meter diameter UKIRT telescope on Mauna Kea, Hawaii. Image credit: Paul Hirst, UKIRT

the northern hemisphere, and a large patch of sky that overlapped with the visible-wavelength Sloan Digital Sky Survey (see section 12.1).

In 2015 the UKIRT telescope was transferred to a new partnership comprising the University of Arizona and the Lockheed-Martin Space Systems Company. They have plans to continue to conduct sky surveys with UKIRT in partnership with astronomers from the UK and University of Hawaii. The resulting survey is called UHS, standing for UKIRT Hemisphere Survey, and will cover the whole sky between declinations $0°$ and $+60°$ that has not already been mapped by UKIDSS. The initial survey is being conducted at a wavelength of $1.25\,\mu m$, but there are plans to extend it to $2.2\,\mu m$.

The European Southern Observatory (ESO) has initiated an infrared survey that is complimentary to UKIDSS in covering the southern hemisphere. It is called VISTA (Visible and Infrared Telescope for Astronomy) and makes use of a new 4.1-meter telescope at Paranal Observatory in Chile. The photographs of UKIRT and VISTA (Figures 6.10 and 6.11) illustrate some of the changes in telescope design that have come about between the completion of UKIRT in 1979 and VISTA in 2009. The most obvious difference is that UKIRT is equatorially-mounted and VISTA has an alt-azimuth mount. The latter design is easier to engineer, but made telescopes much harder to point accurately until the development of reliable computer control.

VISTA will spend about five years observing the whole southern sky to about the same limit as the UHS, but will spend extra time on selected targets such as the Magellanic Clouds. Its 3-tonne camera, built in the UK, contains 67 million pixels.

Fig. 6.11 The 4.1-meter VISTA telescope in Paranal, Chile. Image credit: European Southern Observatory

6.7 SCUBA-2

As indicated in Figure 6.1, there are atmospheric windows at wavelengths of around 350, 450, and 850 μm that can be exploited from ground-based observatories at high, dry sites. Most of the brightest objects in the sky at these wavelengths are molecular clouds in the Milky Way of the kind discussed in section 5.7. Within these clouds the dust particles are heated to temperatures of a few tens of degrees Kelvin. These particles radiate most of their energy at wavelengths that can only be observed from space, but many of them also emit enough radiation in the 350-μm to 1-mm range that they can be mapped from the ground.

The most extensive ground-based surveys at these wavelengths are those conducted by the James Clark Maxwell Telescope (JCMT) on Mauna Kea, Hawaii (Figure 6.12). This 15-meter diameter paraboloid antenna started operations in 1987 as a partnership between the United Kingdom, Canada, the Netherlands, and the University of Hawaii. In 2015 it was transferred to the East Asian Observatory, which is mainly funded by Japan, China, Taiwan, and South Korea.

The JCMT is equipped with a camera called SCUBA-2, which stands for Submillimeter-Common-User-Bolometer-Array. It contains separate 5120-pixel arrays for both 450 and 850 μm wavelengths. It operates in the same wavelength range as the Planck satellite (section 7.4) but since the mirror of the JCMT is ten times the diameter of that in the Planck satellite, it is capable of mapping much finer detail.

Fig. 6.12 The JCMT telescope on Mauna Kea, Hawaii. The telescope is usually protected from the wind by a Gore-Tex membrane which fills the opening of the dome. Image credit: UK Scientific & Technical Facilities Council

The combination of JCMT and SCUBA-2 has been widely used to map many regions of the sky, both in the galactic plane and outside it. The survey that is currently being carried out with these instruments has the acronym SASSy. The acronym originally stood for "SCUBA All Sky Survey" but it was soon realized that the observing time needed to cover the whole sky with good sensitivity was more than could be reasonably accommodated. Now that its planned scope has been somewhat reduced, the acronym stands for "SCUBA Ambitious Sky Survey."

Chapter 7
The Cosmic Microwave Background

The millimeter waveband, between about 0.3 mm and 1 cm (Figure 7.1), has given us wonderful insights into the nature and distribution of interstellar matter, via the carbon monoxide spectral lines (section 5.7) and emission from warm dust grains (section 6.7). But by far its most important contribution to astronomy has been the study of the Cosmic Microwave Background radiation (CMB), a topic that earns its own chapter in this book. An additional reason for giving it a chapter to itself is that our understanding of the CMB draws almost equally on the techniques of radio astronomy and infrared astronomy—specifically the use of heterodyne-type radio receivers (which measure oscillating electric fields) and of bolometer-type infrared detectors (which collect and measure photon energies).

7.1 Discovery of the CMB

The discovery of the CMB runs remarkably parallel to Jansky's discovery of emission from the Milky Way thirty years earlier (section 5.1). In 1964, Arno Penzias and Robert Wilson were working at the Bell Telephone Laboratories in New Jersey attempting to account for electrical noise that was limiting the sensitivity of commercial short-wave radio communications. They used a horn antenna that could be pointed anywhere in the sky (Figure 7.2). After eliminating all other possibilities they concluded that their equipment was receiving faint radio signals from beyond the Galaxy that seemed to be the same in every direction they looked at. They calculated that the radiation corresponded to that from a "black body" surface (see Appendix A.9) at a temperature of around 4 K.

An explanation for this emission was forthcoming almost immediately. In 1948 Ralph Alpher and Robert Herman, following on from earlier ideas by George Gamow, had predicted that radiation of this kind would be naturally produced as a by-product of the Big Bang origin of the universe and calculated that it would have an intensity of around 5 K. According to the now widely-accepted theory, the universe was very hot when it was created, but cooled down as it expanded. At its

© Springer International Publishing Switzerland 2016
G. Wynn-Williams, *Surveying the Skies*, Astronomers' Universe,
DOI 10.1007/978-3-319-28510-8_7

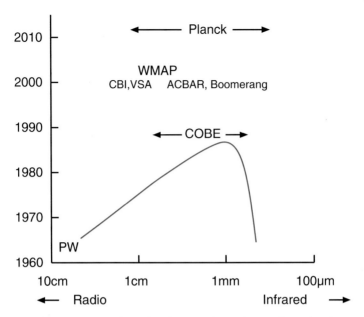

Fig. 7.1 The millimeter waveband showing the approximate dates and wavelength ranges of the surveys discussed in this chapter. The red curve shows the approximate range of the Cosmic Microwave Background

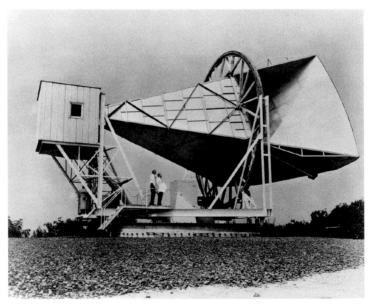

Fig. 7.2 Robert Wilson and Arno Penzias standing under the antenna with which they discovered the cosmic background radiation. Image credit: Wikipedia

initial high temperature the universe was entirely a plasma with all its atoms ionized, but after about 380,000 years it had cooled to a temperature of about 3,000 K and thinned enough so that protons and electrons, the most abundant particles, were able to combine to form neutral hydrogen atoms. When this happened the universe quickly became transparent. Ultraviolet photons that had been bouncing from one charged particle to another suddenly found themselves able to pass through the universe unimpeded. As the universe expanded so did the wavelength of these photons; at the present time they correspond to the radiation expected from a black body at a temperature of a few degrees Kelvin, generating emission that peaks near a wavelength of 1 mm, extending into both the radio and far-infrared wavebands. This was the radiation that Penzias and Wilson had discovered.

7.2 COBE

Penzias and Wilson's original observations, which earned them a Nobel Prize in 1978, were made at a wavelength of 7.35 cm. Follow-up measurements at other radio wavelengths confirmed that the emission grew stronger at shorter wavelengths and that its strength corresponded to a brightness temperature of 2.7 K. But to confirm that the emission did indeed correspond to a theoretical black body it was necessary to push observations to wavelengths of less than 1 mm, where theory predicted that the radiation would peak and then drop off (see Figure 7.1). This requirement led to the design and construction of the Cosmic Background Explorer (COBE), a NASA satellite that was launched in 1989 (Figure 7.3).

Fig. 7.3 Artist's impression of the COBE satellite in orbit. The telescope always points away from the Earth. Image credit: NASA/COBE Science Team

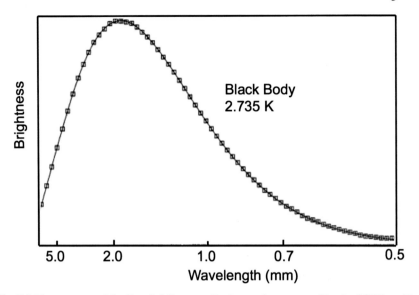

Fig. 7.4 The spectrum of the Cosmic Microwave Background as measured by the COBE satellite, illustrating how well the data points fit the theoretical curve for a black body. Based on an image by NASA/COBE Science Team

COBE was fitted with special instruments that measured the strength of the cosmic background radiation in the millimeter waveband. The analysis of the data required careful subtraction of the effects of warm dust both within the solar system and in the Galaxy. The final result was an almost perfect fit between the measured CMB and the calculated black-body emission of an object at a temperature of 2.735 K (Figure 7.4) The agreement with theory provided very strong support for the Big Bang theory and helped turn cosmology into a precise science.

At the time that the CMB was generated, 380,000 years after the Big Bang, the gas in the universe was beginning to clump together into the precursors of what would later become galaxies and clusters of galaxies. Cosmologists calculated that this clumping would produce minute, but measurable fluctuations in the strength of the CMB in different regions of the sky; searching and measuring these fluctuations has since been a major challenge for astronomers. The first instrument to detect CMB fluctuations was the COBE satellite, which carried a mapping instrument as well as the spectrometer that was used to make the measurements shown in Figure 7.4. After subtracting off the emission from foreground sources such as the Milky Way galaxy, the COBE team found random fluctuations of about one part in 100,000 in the strength of the CMB from different regions of the sky. COBE's success led to the 2006 Physics Nobel Prize being awarded to its two leading scientists; George Smoot who was responsible for the measuring the CMB spectrum in Figure 7.4 and John Mather, who was responsible for discovering the fluctuations.

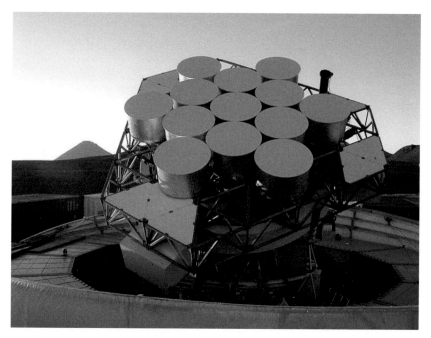

Fig. 7.5 The Cosmic Background Imager at the Chajnantor Observatory at an altitude of 5080 m in the Chilean Andes. Image credit: Caltech

7.3 Mapping the Fluctuations in the CMB

The discovery by COBE of fluctuations in the CMB (sometimes referred to as anisotropies) set off a frenzy of excitement among astronomers in the 1990s and 2000s, and inspired the construction of a number of telescopes for mapping the CMB in finer detail than COBE had been able to do. Most of these telescopes were built at high, dry sites where contamination by water vapor in the Earth's atmosphere is minimized. Among them were the Very Small Array (VSA) in Tenerife, the Cosmic Background Imager (CBI) in Chile (Figure 7.5), and the Arcminute Cosmology Bolometer Array Receiver (ACBAR) at the South Pole. The first two instruments used aperture synthesis radio astronomy techniques to map selected regions of the sky in the wavelength range 8–12 mm. Acbar used an array of solid-state infrared detectors to collect radiation at around 1 mm wavelength.

A different approach was taken by the Boomerang experiment, standing for Balloon Observations of Millimetric Extragalactic Radiation and Geophysics. This consisted of a 1.2-meter diameter telescope suspended below a helium-filled balloon that was launched from the McMurdo base in Antarctica (Figure 7.6). This site had an enormous advantage over any launch site at more northern latitudes because the prevailing winds carried it in a 8,000 kilometer circular journey round the South Pole which lasted 10 days and ended up only 50 kilometers from where it had started.

Fig. 7.6 The Boomerang telescope being launched for its 10-day trip around the South Pole. Image credit: Wikipedia

Boomerang used bolometers rather than radio receivers, so was able to observe at wavelengths in the 0.8–3.3 mm range where the CMB is at its strongest.

None of these experiments could map fluctuations over the whole sky, so in 2001 NASA launched the WMAP satellite which stands for "Wilkinson Microwave Anisotropy Probe." It was active until 2010. In order to minimize the effects of scattered radiation from Earth, WMAP was put into orbit around the Sun near what is called the second Lagrangian point, or "L2". This location is named after the Italian mathematician and astronomer Joseph-Louis Lagrange (1736–1813). Lagrange showed that if one considered the combined gravitational forces of the Earth and the Sun there were five places (labelled L1–L5 in Figure 7.7) where an object would orbit the Sun at a fixed distance from both the Sun and the Earth. The most useful of these points for most astronomical purposes is L2, which lies about 1.5 million km from the Earth (nearly four times the distance to the Moon) on the side of the Earth farthest from the Sun. A satellite at this location, when looking in a direction away from the Sun and the Earth, has an extremely stable environment, with no daily or monthly variations in the radiation it receives.

Figure 7.8 shows the map of the CMB produced by WMAP. The variations in temperature are very greatly exaggerated; the difference in brightness temperature between the red and the dark blue areas is only 0.0002 K.

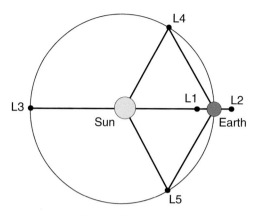

Fig. 7.7 The five Lagrangian points of the Sun-Earth system

Figure 7.9 summarizes the results of these five studies. It is a statistical analysis of the strengths of the brightness variations as a function of their angular scale on the sky: large scale variations are on the left, and fine-scale variations are on the right. The largest peak shows that the dominant variations are about 1 degree in size, with several subsidiary peaks on a finer angular scale. The large-scale data to the left of the picture comes from WMAP; the small-scale data on the right are derived from the ground- and balloon-based studies.

The full interpretation of this graph requires a detailed understanding of relativistic cosmology, but the basic idea is fairly simple. During the first 380,000 years of the universe, gravitational attractions within the ionized gases resulted in the formation of concentrations whose sizes depended on such factors as the pressure

Fig. 7.8 Map of the CMB produced by the WMAP satellite survey. Image credit: NASA/WMAP Science Team

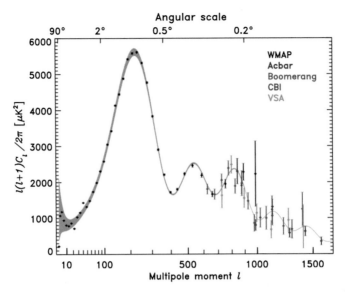

Fig. 7.9 Fluctuations in the Cosmic Microwave Background. Image credit: NASA/WMAP Science Team

and the speed of sound in the gas as well as on the gravitational forces from both ordinary matter and dark matter. Other phenomena such as the Doppler effect and the gravitational bending of light also played a role in generating fluctuations on different scales. By performing a statistical analysis of these fluctuations cosmologists were able to deduce an extremely precise value for the age of the universe $(13.77 \pm 0.05$ billion years). They also determined that ordinary matter (such as protons, neutrons and electrons) comprises only 4.6 % of the total amount of matter in the universe, the remainder being comprised of dark matter and dark energy. They were also able to conclude that the universe will expand forever. The precision of these results is remarkable, not least because as recently as the 1970s astronomers could not agree on the value of the age of the universe to within a factor of two.

7.4 The Planck Spacecraft

The most recent and most comprehensive study of the CMB was made by the Planck satellite, which was launched by the European Space Agency in 2009 and operated until 2013. It was placed into an L2 orbit similar to that of WMAP, but outperformed WMAP in several ways, notably in its ability to map finer details, and in its much wider range of observing wavelengths. Planck used radio receivers for the longer wavelength bands (3.9–11 mm) and solid state infrared detectors for the shorter wavelengths (0.3–3.6 mm); the latter were cooled to within 0.1° of absolute zero.

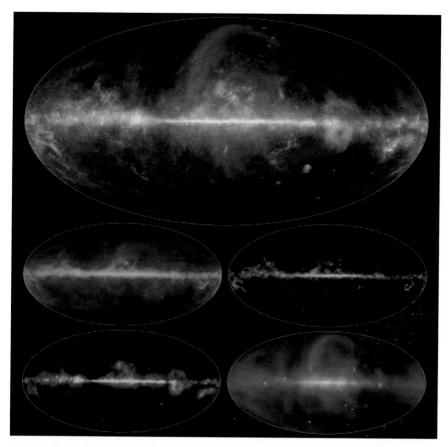

Fig. 7.10 Emission from the Milky Way galaxy as measured by the Planck spacecraft. Red color indicates emission from warm dust grains, yellow shows carbon monoxide gas, blue represents synchrotron radiation, and green is radiation from hot ionized gases such as those in the Orion Nebula. The top map is a combination of these radiation sources. Image credit: ESA/NASA/JPL-Caltech

The wide range of observing wavelengths helped the Planck astronomers separate radiation from the cosmic background from spurious foreground radiation generated from within our Galaxy. Guided by existing maps at longer and shorter wavelengths, the Planck scientists were able to produce a composite picture of the mm-wave emission from our Galaxy (Figure 7.10). Once the emission from these sources of galactic radiation had been eliminated, Planck was able to produce the most detailed map of the CMB ever produced (Figure 7.11). As can be seen in Figure 7.12 the level of detail is far better than that provided by COBE and significantly better than that obtained from WMAP.

Fig. 7.11 The CMB as mapped by the Planck satellite. Image credit: NASA/JPL-Caltech/ESA

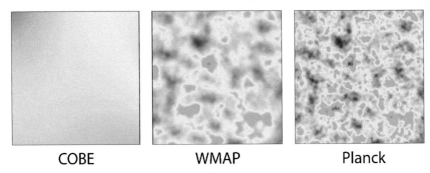

COBE   WMAP   Planck

Fig. 7.12 The same 10 square degree patch of sky as seen by the COBE, WMAP, and Planck satellites. Image credit: ESA/NASA/JPL-Caltech

Careful statistical analysis of the scale sizes of the fluctuations led to even more precise estimates of the age of the universe than revealed by WMAP. It also provided strong support for the current mainstream cosmological theories that include the idea that the universe is geometrically "flat" and that it will expand forever. Astronomers had hoped to see evidence for gravitational waves in the very early universe, because cosmologists had predicted that they would produce a particular pattern of electromagnetic polarization in the CMB maps. Intense efforts were made to detect this effect, but it has so far been masked by the polarized radiation generated by spinning dust particles in the local interstellar medium of the Milky Way. The Planck mission will therefore probably not be the last word on CMB research.

Chapter 8
Ultraviolet Surveys

The discovery of ultraviolet radiation in 1801 followed very closely after the discovery of infrared radiation. On hearing of Herschel's experiments, the German scientist Johann Ritter (1776–1810) searched for corresponding signs of radiation beyond the violet end of the Sun's spectrum. He already knew that silver chloride slowly turns black in the presence of light, so he decided to measure the strength of this effect at different places in the spectrum. He found that the effect got stronger towards the violet end of the spectrum and peaked where the radiation was invisible. Ritter's name for this radiation was Chemical Rays; the term ultraviolet came later.

The long-wavelength limit of the ultraviolet waveband is generally considered to be 0.4 μm, which is the same as 400 nanometers (nm). A small range of wavelengths, roughly 300–400 nm, can get through the atmosphere and tan our skin, but it is not picked up by our eyes: this range is sometimes called the photographic ultraviolet since photographic plates used to be available that were sensitive in this waveband. Ultraviolet studies at wavelengths shortward of 300 nm have to be done from space, and it is with surveys at these wavelengths that this chapter is concerned (see Figure 8.1). Wavelengths shorter than 10 nm are usually referred to as X-rays, though the boundary between ultraviolet and X-ray is not strictly defined.

It is not only the Earth's atmosphere that hinders ultraviolet astronomy: the interstellar medium is a major problem as well. Extinction due to interstellar dust grains rises steadily as one moves to shorter ultraviolet wavelengths, but at 91 nm a second effect becomes important, namely the ionization of interstellar neutral hydrogen atoms. Photons with wavelengths less than 91 nm (which is sometimes referred to as the Lyman limit) have enough energy to split a neutral hydrogen atom into a positively-charged proton and a negatively charged electron. These two particles will then travel separately around interstellar space until they happen to meet another example of their opposite number, whereupon they will recombine to form a neutral hydrogen particle plus another ultraviolet photon. The new ultraviolet photon is very unlikely to travel in the same direction as the photon that was absorbed, so the overall effect is that ultraviolet light traveling through this region of space is scattered by the hydrogen. Only if the region of space in question is being

© Springer International Publishing Switzerland 2016
G. Wynn-Williams, *Surveying the Skies,* Astronomers' Universe,
DOI 10.1007/978-3-319-28510-8_8

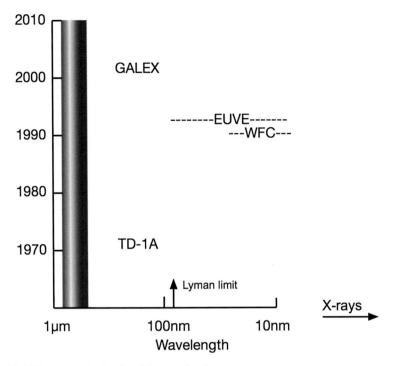

Fig. 8.1 Major surveys in the ultraviolet waveband

kept in an ionized state by some other means will photons with wavelength less than 91 nm travel easily through it.

Because of the jump in interstellar extinction at the Lyman limit, ultraviolet surveys divide themselves naturally into two categories: "mid ultraviolet" surveys such as TD-1A and GALEX, which are made at wavelengths longward of 91 nm, and "extreme ultraviolet" surveys such as WFC and EUVE, which are performed in the range between 91 nm and the X-ray region.

The comparatively small number of sky surveys discussed in this chapter gives a misleading impression of the scientific importance of ultraviolet astronomy. Most of the elements found in the atmospheres of stars exhibit more spectral lines in the ultraviolet region than they do in the visible region; for this reason there has been a series of highly successful spacecraft devoted to studying the spectra of already-known stars. Prime examples of these are the Copernicus satellite launched in 1972, the International Ultraviolet Explorer (IUE) launched in 1978, and the Hubble Space Telescope launched in 1990. The ultraviolet is particularly important for the study of the hottest kinds of stars, since they emit most of their energy in that region. Such stars include white dwarfs—the endpoints of stars like the Sun—and main-sequence OB stars, the powerhouses of HII regions like the Orion Nebula.

8.1 TD-1A

The first space-borne survey of the sky at ultraviolet wavelengths was the TD-1A satellite, launched into orbit around the Earth by the European Space Agency in 1972. It was a comparatively small satellite with a mixture of X-ray and UV experiments. The telescope always pointed directly away from the Earth and swept the sky continuously during its 90-minute orbit. The orbit of the satellite was designed so that after six months its telescope would have pointed everywhere in the sky at least once. There was no need to control the telescope's pointing once it was in the correct orbit.

The ultraviolet survey observations were made with a 27.5 cm telescope. A diffraction grating was used to form the spectrum of each star as it passed in front of the telescope. The resulting signals were collected with photomultipliers and radioed back to Earth. Continuous spectra covering the range 130–260 nm were obtained in this way for about 1800 bright stars. In the case of fainter stars the spectra were averaged to improve their signal to noise ratio. The final TD-1A catalog contains flux density measurements at four ultraviolet wavelengths for 31,215 objects. All but 10 of the objects discovered by the TD-1A survey were identified with stars that were already known, numbered and, in most cases, spectrally classified. As expected, most of these stars are of spectral types O, B, A, or F, meaning they have higher surface temperatures than the Sun.

Figure 8.2 shows a map of the stars detected by TD-1A. Although there is some horizontal concentration of stars, the Milky Way is far less prominent at ultraviolet wavelengths than at infrared wavelengths (see Figures 6.3 and 6.8). The reason for this is that the strong interstellar extinction at ultraviolet wavelengths prevents

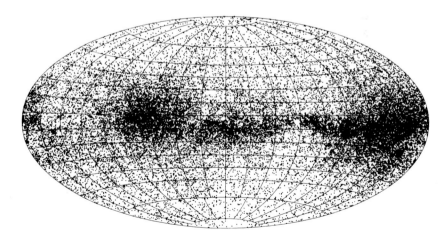

Fig. 8.2 Map of stars detected in the TD-1A ultraviolet survey in galactic coordinates. Image credit: European Space Research Organization

us from seeing more than a few hundred parsecs through the disk of the Milky Way, whereas at infrared wavelengths we can see as far as the Galactic Center and beyond. The main concentrations we see in Figure 8.2 are towards the constellations Cygnus (to the left of center) and Orion (at the right edge of the map). These are the approximate directions of our local spiral arm.

8.2 GALEX

Besides TD-1A, the only other all-sky survey in the 100–300 nm wavelength range was NASA's Galaxy Evolution Explorer (GALEX) mission, which was launched in 2003 and sent data back to Earth until 2013. Despite the fact that the telescopes themselves differed in diameter by a factor of less than two, technical developments over the intervening 30 years gave GALEX enormous advantages over TD-1A both in resolution and sensitivity (Table 8.1). GALEX could detect sources that were 10 magnitudes (a factor of 10,000) fainter than TD-1A, and resolve details that were hundreds of times finer.

The two telescopes operated quite differently. Whereas TD-1A was left to itself to scan the sky in an unwavering regular spiral pattern, GALEX was sent commands telling it to point accurately in particular directions in the sky and then record images with typical exposures of 100 seconds. Each exposure had a field of view of 1.2° so that around 27,000 images had to be recorded to view the whole sky. Certain regions of particular interest were observed for much longer, and in these special areas the detection limits were up to 100 times fainter that for the full survey. On the other hand, the detector system was so sensitive that a few areas of the sky, including much of the galactic plane, had to be avoided because of potential overexposure.

As its name implies, GALEX was built to study galaxies, and it produced some very beautiful images of both nearby and distant galaxies. The GALEX images are particularly useful for pinpointing the regions of a galaxy where new stars are being formed. The reason for this is that the main-sequence stars that produce the most ultraviolet light—spectral types O and B—live for only a comparatively short time before evolving into red giant stars and supernovae. Their lifetimes are too short for them to stray very far from the region of the galaxy in which they were born, so they can act as markers for the places where stars formed within the last few million years. As an example, Figure 8.3 shows the difference between the appearance of

Survey name	TD-1A	GALEX
Observing Epoch	1972–1974	2003–2013
Telescope Diameter (m)	0.27	0.5
Number of detectors	4	~8,000,000
Beamsize (arcseconds)	700×1000	4×4

Table 8.1 Comparison of the TD-1A and GALEX all-sky ultraviolet surveys

Fig. 8.3 The spiral galaxy M81 as seen by visible light (*left*) and ultraviolet light (*right*). See Figure 5.12 for a map of the 21-cm line emission from this galaxy. Image credit: GALEX NASA/JPL-Caltech/NOAO

the galaxy M81 at visible and ultraviolet wavelengths. It shows that most of the new stars that are forming in this galaxy are in its outer spiral arms rather than in the central bulge.

GALEX did not spend all its life making an all-sky survey. It was a major contributor to the Cosmic Galaxy Evolution Survey (COSMOS), which involved a detailed multi-wavelength study of a $2° \times 2°$ field. The COSMOS survey is described in section 13.2.

8.3 Extreme Ultraviolet Surveys

There have been two major sky surveys in the extreme ultraviolet waveband short-ward of the Lyman limit at 91 nm. These were the Wide Field Camera (WFC) on the ROSAT satellite, and the Extreme Ultraviolet Explorer satellite (EUVE). Both surveys were made during the 1990s.

ROSAT was built and operated by the German Aerospace Center. As we shall discuss in Chapter 9, most of its instruments were designed to observe at X-ray wavelengths, but the WFC collected radiation in two wavebands between 6 and 20 nm. The telescope used aluminum mirrors to focus the radiation onto an array detector of 512×512 pixels, giving a spatial resolution of about 2 arcminutes over a $1°$ field of view. The satellite was launched in 1990 and placed into an orbit roughly similar to that of TD-1A, so that the whole sky could be observed in six months. The final catalog of the WFC survey contained 479 objects fairly evenly spread around the sky.

The EUVE satellite was launched by NASA in 1992, two years after ROSAT. It had a wider wavelength converge than WFC, allowing it to observe out to 80 nm wavelength. The satellite spent much of its first year in orbit conducting an all-sky survey; when this was complete it spent several years making more detailed studies of objects specifically chosen by astronomers, and conducting a deep survey of a particular region of sky. The final all-sky survey catalog contained 287 sources.

There is good agreement between the catalogs produced by the two surveys. Roughly 40 % can be identified with nearby ordinary stars with spectral types FGK or M and 30 % with white dwarf stars. A total of 21 objects were identified with objects outside the Milky Way. Most of them are quasars or other active galaxies, and all have independently been established to be bright sources of X-ray emission.

At first sight, the presence of so many extragalactic objects is a surprise, given the high opacity of the interstellar medium in the extreme ultraviolet, but a more detailed analysis showed that all the extragalactic objects were in one of two areas in the sky, both well away from the plane of the Milky Way. By combining these results with spectroscopic observations of many of the stars found in the survey, astronomers have been putting together a new picture of the interstellar medium within about 200 parsecs from the Sun (Figure 8.4). The Sun, and most of its neighbors within 100 parsecs, lies in a region where the interstellar density is unusually low—

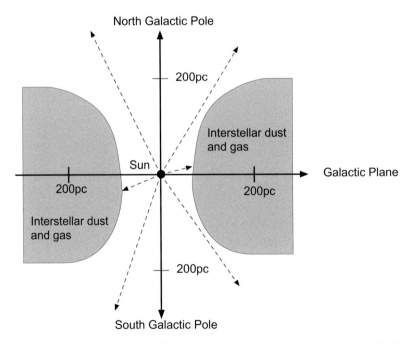

Fig. 8.4 Sketch of the vicinity of the Sun as revealed by ultraviolet observations. In this picture the plane of the Milky Way is horizontal and we are facing away from the galactic center. The Sun and most of its neighbors within 100 parsecs lie within the "Interstellar Chimney" of low density matter, that extends upwards and downwards into intergalactic space

probably as the result of a prehistoric supernova explosion somewhere in this region. The shape of the low density area is roughly tubular, extending all the way through the disk of the Milky Way. It has therefore become known as the "Interstellar Chimney." All the extragalactic objects observed by WFC and EUVE lie either above or below the galactic plane in the direction of the chimney. Present day astronomers can be thankful for the supernova or group of supernovae that generated the chimney; without it we would know much less about the ultraviolet properties of distant galaxies.

A problem with both the WFC and EUVE surveys was that each object in the sky was observed very few times, so that variable sources might be anomalously bright or faint when they happened to be observed. The ALEXIS survey, launched in 1993, used a much smaller satellite and had much poorer spatial resolution than WFC or EUVE, but it surveyed the whole sky much more often than did the larger satellites. Its main success was the discovery and study of several highly variable far-ultraviolet sources. Three of these could be identified with unusual binary star systems, but two others have yet to be reliably identified.

Chapter 9
X-ray Surveys

X-rays were discovered by accident in 1895 by Wilhelm Röntgen who was studying electrical discharges in gases. His work won him the first-ever Nobel Prize for physics in 1901.

X-rays are generally defined as having wavelengths less than about 12 nm. However, the convention among astronomers and physicists is to define X-rays in terms of their photon energies rather than their wavelengths (see Appendix A.3). Under this convention the X-ray spectrum starts at an energy of around 0.1 keV and extends up to about 100 keV, where the realm of γ-ray astronomy begins. Photons on the low-energy end of this range (0.1–10 keV) are sometimes referred to as "soft X-rays" while those at the high-energy end (10–100 keV) are referred to as "hard X-rays," but these boundaries are not at all well-defined.

At first sight, the X-ray band is not a promising one for astronomy. Wien's Law (Appendix A.9) implies that for an object to radiate strongly at a wavelength of 1 nm (1.2 keV) it would need to have a temperature of more than 1,000,000 K. An object of stellar size at this temperature would radiate energy so quickly that it would rapidly cool and fade from view. Fortunately for astronomy, some pioneers in the subject chose to ignore this problem and search for celestial X-ray sources anyway. As we shall see in this chapter, their optimism was more than justified.

9.1 Early Discoveries

The Earth's atmosphere is opaque to X-rays, even above the highest mountains. Apart from a few specialized experiments carried under high-altitude balloons, all X-ray astronomy has been done from space using either rockets or satellites.

A group led by Herbert Friedman at the US Naval Research Laboratory first detected X-rays from the Sun in 1948 using a captured German V-2 rocket launched from White Sands Missile Range in New Mexico. The V-2 rockets were soon replaced by US-designed ones (Figure 9.2), but the overall observing pattern remained much the same; the X-ray telescope system was launched more or less vertically

© Springer International Publishing Switzerland 2016
G. Wynn-Williams, *Surveying the Skies*, Astronomers' Universe,
DOI 10.1007/978-3-319-28510-8_9

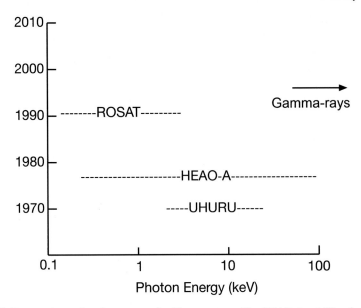

Fig. 9.1 Dates and wavelength coverage for X-ray surveys. The HEAO-A satellite also carried gamma-ray experiments. To relate this figure with Figure 8.1 note that 0.1 keV is equivalent to 12 nm (see Appendix A.3)

with enough speed that it would spend several minutes above the atmosphere moving first upwards and then downwards before a parachute opened to bring the telescope X-ray detector system back to Earth. The rockets themselves crashed and were destroyed.

Rocket-based X-ray studies of the Sun continued during the 1950s, but during this time there was little optimism about detecting any other stars, given their far greater distance from the Earth. In 1962, however, a group of scientists led by Riccardo Giacconi detected a second X-ray source in the sky which, to their great surprise, outshone the disk of the Milky Way, but did not coincide with any bright star (Figure 9.3). It was obvious that they had discovered a new kind of celestial object; they named it Sco X-1 after the Scorpius constellation in which it was found. Follow-up studies showed that Sco X-1 could be identified with a 13th magnitude blue star whose brightness fluctuated in ways that were reminiscent of a nova.

The following year the NRL team detected X-ray emission from the Crab Nebula supernova remnant—the first X-ray source to be identified with a previously-studied object. Follow-up observations showed that the X-rays come primarily from the extended nebula rather than a compact source and that they might be caused by the same synchrotron processes that causes the Crab's radio emission.

Fig. 9.2 Early X-ray astronomy rocket being prepared for launch. Image credit: Harvard Center for Astronomy, High Energy Astrophysics Division

By 1969 about 30 X-ray sources had been discovered during dozens of rocket flights, despite the fact that the total amount of observing time amounted to only about 2 hours over 20 years. They included the radio galaxy M87, the quasar 3C273, a few supernova remnants, and a number of sources whose X-ray flux was highly variable. The available evidence strongly suggested that many of these X-ray sources were connected with highly unusual, sometimes binary stars. There was also evidence for a diffuse background of X-ray emission from the whole sky. X-ray astronomy had quickly earned itself a secure position in the frontline of astronomical research.

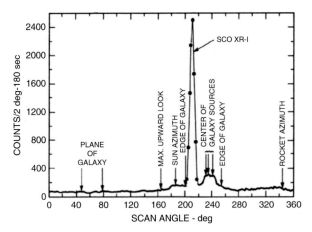

Fig. 9.3 Three minutes of a scan from a rocket-borne X-ray detector made in 1967 showing how the source Sco X-1 stands out above the emission from the Milky Way. ©American Astronomical Society, from ApJ 154, 655 (1968)

9.2 UHURU and HEAO-A

The successes of the rocket program naturally led to calls for the development of a satellite that could provide months instead of minutes of observing time, could survey the whole sky at X-ray wavelengths, could perform detailed studies of selected regions, and could monitor time-varying sources for days at a time. Such a satellite, originally named SAS-A, was built and ready for launch in 1970.

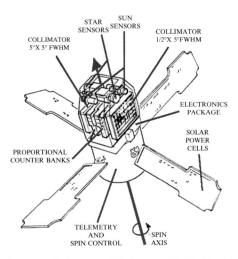

Fig. 9.4 Diagram showing the main features of the UHURU (SAS-A) satellite. The overall height is about 2 meters. Image credit: High Energy Division, Smithsonian Astrophysical Observatory

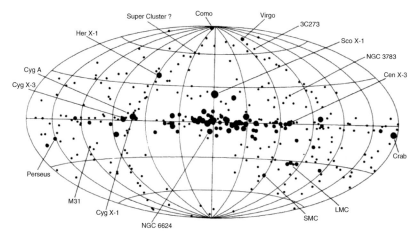

Fig. 9.5 X-ray Map of the sky in galactic coordinates produced by the UHURU satellite. ©American Astronomical Society, from ApJSS 38, 357 (1978)

One problem affecting X-ray astronomy is that charged particles high in the Earth's atmosphere can produce spurious signals in the detectors. The degree of interference that these particles produce depends on the geographic latitude of the satellite, and is worst near the Earth's poles. To minimize their effects the satellite was therefore launched eastwards from a retired oil-drilling platform a few kilometers offshore from the coast of Kenya in east Africa, and placed into an orbit more or less directly above the Earth's equator. The platform had been put there by the Italian Space Agency for their own satellite program. After SAS-A was successfully launched it was renamed UHURU, which is Swahili for freedom.

Strictly speaking, neither the UHURU satellite nor the rocket experiments that preceded it contained a telescope, in the sense of an instrument that focuses radiation onto a detector. The main instrument was a proportional counter, which is a device that converts an incoming X-ray photon into a brief electrical current whose intensity is a measure of the photon's energy. The instrument was sensitive to X-rays in the energy range 2–20 keV. Collimating slats placed outside the proportional counter restricted its view to one particular direction (Figure 9.4). The sky was scanned by letting the satellite spin at a rate of once every 12 minutes. Two collimator systems were used: one limited the size of the patch of sky under study to 5° × 5° the other to 5° × 0.5°, which meant that objects in the sky could have their positions measured to a accuracy of a few minutes of arc.

During its 2.5 year lifetime in orbit, UHURU scanned 95 % of the sky leading to a final published catalog of 339 sources (see Figure 9.5). The faintest of these was about 1000 times fainter than Crab Nebula. Many of the X-ray sources that UHURU discovered could be identified with already known objects, such as normal galaxies, Seyfert galaxies, supernova remnants, and clusters of galaxies. But it also discovered many new objects, including the X-ray binary systems we will discuss in the next section.

The great success of the UHURU mission led to a proliferation of X-ray astronomy satellites from several different countries. Most of them were designed to make detailed studies of already-known X-ray sources rather than survey the sky for new ones. Seven years after UHURU's launch, however, NASA launched HEAO-A, which was the first in a series of three High Energy Astrophysics Observatories. It included two X-ray survey instruments, which scanned the whole sky over a period of more than a year. One of the surveys led to a catalog of 842 objects detected in the energy range 0.25–25 keV, while the other found 70 sources in the energy range 13–80 keV. The majority of them were in the plane of the Milky Way galaxy,

9.3 Neutron Stars and Black Holes

The UHURU satellite did not spend all of its time surveying the sky. At the time of its launch there was already good evidence that some X-ray sources were variable, but methodical study of these variations was almost impossible in the era of 5-minute duration rocket flights. UHURU therefore spent part of its time monitoring the X-ray emission from a few already-known sources.

The first breakthrough of the program was the discovery that the X-ray source Centaurus X-3 showed periodic variations of about 5 seconds, reminiscent of the behavior of the recently discovered radio pulsars. Small, but regular variations in the 5-s period indicated that the pulsating X-ray source was part of an eclipsing binary star system with a period of 2.1 days. With this information optical astronomers searched the patch of sky and soon found a previously unnoticed massive blue star with the same binary period. Soon after this, another X-ray source, Hercules X-1, was found with a pulse period of 1.24 s and an orbital period of 1.7 days.

A natural explanation for these binary systems emerged quickly: they consisted of a compact, rapidly rotating neutron star partnered with a larger, more conventional star (Figure 9.6). The two stars are so close that gas is pulled away by gravity from the surface of the large star towards the neutron star, which has a diameter of only about 20 kilometers. As it rushes towards the neutron star the gas gathers kinetic energy which raises its temperature to around a million degrees. Because the whole binary system is rotating, the in-falling gas forms what is called an "accretion disk" around the neutron star, and it is this hot disk that is the origin of most of the X-ray emission from the system.

Neutron stars result from supernova explosions, which occur when a large, normal star has exhausted all its potential nuclear fuel and is no longer able to support itself by internal pressure. The central part of the star suddenly collapses in on itself releasing an enormous amount of gravitational energy, which then blows the outer parts of the star into space. Under the right circumstances, the dense core that remains is a neutron star, an object whose density is comparable to that inside an atomic nucleus.

Theoretical calculations show that neutron stars can only have masses of between 1.4 and 3 times the mass of our Sun. If the core of a star has a mass less than 1.4

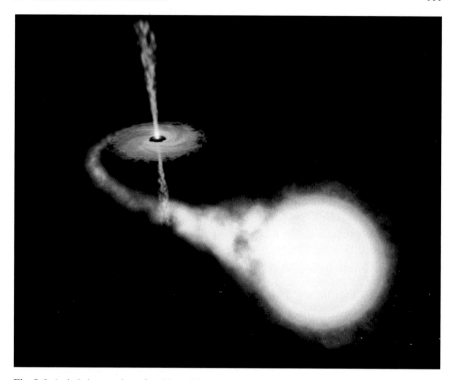

Fig. 9.6 Artist's impression of an X-ray binary system. The neutron star is on the left, surrounded by an accretion disk made up of matter that has been pulled off the companion star on the right. Image credit: European Space agency and NASA

solar masses it collapses only as far as a white dwarf star and no supernova explosion occurs. Conversely, if the mass of the core is greater than 3 solar masses it collapses all the way to a black hole. Because a black hole emits no radiation it cannot be seen directly, but a black hole in a close binary star system can be surrounded by a hot accretion disk in the same way as a neutron star (Figure 9.6).

Astronomers realized that if they could find an X-ray binary system which contained a compact object with a mass greater than 3 solar masses then they could be confident they had discovered a black hole for the first time. Attention soon focussed on the bright X-ray binary star system Cygnus X-1, which was found to coincide with a supergiant B star orbiting a hidden companion star every 5.6 days. Calculations indicated that the hidden companion had to have a mass greater than 3 solar masses, and probably closer to 10 solar masses: the first black hole was thus discovered within 2 years of the launch of UHURU. Others soon followed.

Another famous X-ray source, the Crab Nebula, is also powered by a neutron star, but the mechanism is quite different. At its center is a rapidly spinning neutron star whose magnetic field accelerates electrons to close to the speed of light, enabling them to produce the extended region of synchrotron radiation that dominates the emission at both radio and X-ray wavelengths.

9.4 ROSAT

By far the most comprehensive X-ray all-sky survey was the one performed by the ROSAT satellite which was launched in 1990 and remained in operation for over eight years. It was built by the German Aerospace Center with substantial contributions from the USA and the UK.

ROSAT contained a true X-ray telescope, in the sense of an instrument that can focus electromagnetic radiation onto a focal plane, but the arrangement of its mirrors was very different from that of a visible-light telescope. Most astronomy mirrors are designed to deal with radiation that shines more or less at right angles to the mirror surface. Mirrors like this cannot be used for X-rays, however, because X-ray photons— have enough energy to penetrate the mirror surface rather than be reflected off it. But if an X-ray photon hits a polished metal surface at a very shallow angle it does get reflected, and this property can be used to build what is called a Wolter telescope (Figure 9.7), named after its inventor Hans Wolter (1911–1978). A Wolter telescope resembles a conventional Cassegrain telescope in that the primary mirror is parabolic and the secondary mirror is hyperbolic, but because the photons have to hit the mirrors at angles of less than about $2°$ the mirror surfaces end up looking much more like tubes than disks. In the case of ROSAT, four pairs of concentric parabolic/hyperbolic tubular mirrors were combined into the same instrument, thereby increasing the collecting area of the telescope. The whole optical system was good enough to discern details as small as 5 arcseconds across.

ROSAT was not the first X-ray satellite to use a Wolter telescope—the very successful Einstein Observatory preceded it—but it was the first all-sky survey carried out with such an instrument. Its X-ray observations were made in the energy range 0.1 to 2.4 keV, but it also carried the far ultraviolet Wide Field Camera (WFC) that we discussed in section 8.3.

ROSAT spent its first year in orbit making a map of the whole sky. After this all-sky survey was finished, ROSAT was used to study particular regions of interest, resulting in the discovery of many additional fainter objects. All in all, ROSAT detected more than 150,000 X-ray sources, raising the total number of sources known in the sky by a factor of more than 20. ROSAT was sensitive enough to pick up the X-ray emission from comets, stars, white dwarfs, cataclysmic variables, neutron stars, black hole candidates, supernova remnants, nearby galaxies, active galaxy nuclei, and clusters of galaxies. Astronomers now routinely check the ROSAT catalog for information about the objects they are studying, even if their X-ray emission is not their main point of interest. More than 3,000 scientific papers have been published using ROSAT data.

Figure 9.8 shows maps of the whole sky at three different X-ray energy bands. At the lowest energy (upper figure) the strongest emission comes from the poles of the Galaxy. This is a consequence of the interstellar chimney effect that is illustrated in Figure 8.4; low-energy X-rays from sources in the galactic plane are mainly hidden from us by the interstellar medium, but X-rays from above and below the plane of

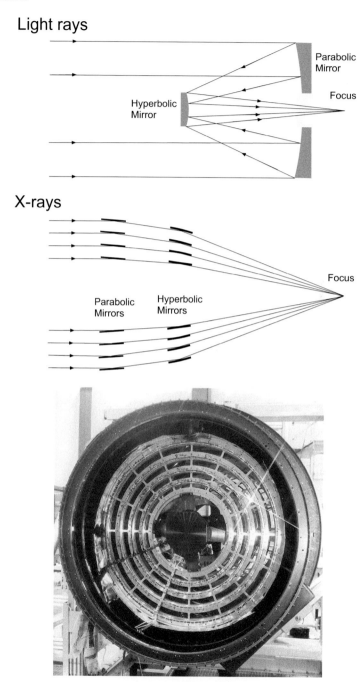

Fig. 9.7 Comparison between the optics of a conventional Cassegrain optical telescope (*top*) and a Wolter X-ray telescope (*middle*). The bottom image shows a face-on view of the 83-cm diameter concentric parabolic mirrors in ROSAT. Bottom image credit: Max-Planck-Institut für Extraterrestrische Physik

1/4 keV

3/4 keV

1.5 keV

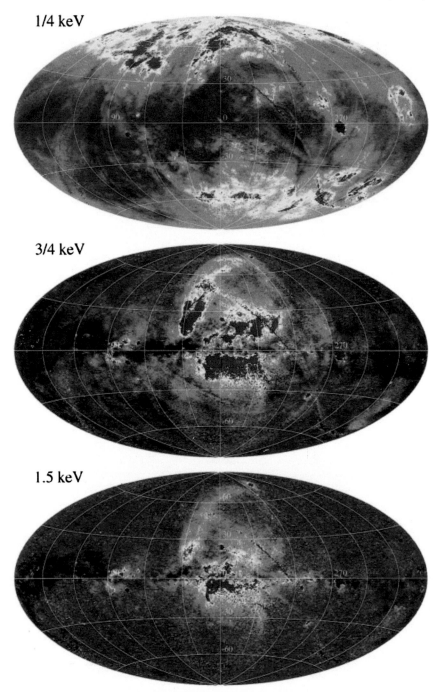

Fig. 9.8 ROSAT maps of the X-ray sky at three different photon energies. Brightest regions are red. The galactic plane runs horizontally across the maps. ©American Astronomical Society, from ApJ 285, 125 (1997)

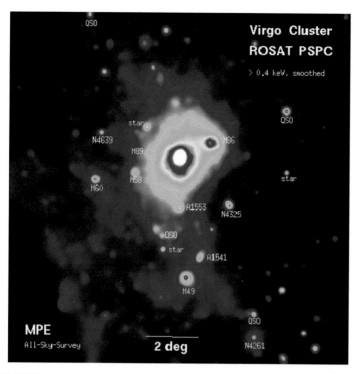

Fig. 9.9 ROSAT map of the Virgo cluster of galaxies, some 16 Mpc away. The brightest region is centered on the giant elliptical galaxy M87. Most of the other compact sources are background quasars unrelated to the cluster. Image credit: Max-Planck-Institut fur extraterrestrische Physik

our Galaxy reach us with less problem. At the higher energies, where the interstellar medium is more transparent, emission from objects within our Galaxy dominate.

At the highest energies, where interstellar extinction is least, ROSAT discovered what is sometimes described as a background of X-ray emission from all directions in the sky. ROSAT showed that most of the background comes from individual active galaxy nuclei, but some of it comes from hot gas in clusters of galaxies. Figure 9.9 shows the nearest of these, the Virgo Cluster, which is several degrees across in the sky. The emission comes from gas that became heated to millions of degrees as a result of being gravitationally sucked into the cluster.

The small number of X-ray sky surveys mentioned in this chapter might give a misleading impression of the current state of X-ray astronomy. During the last 50 years there have been many other highly sophisticated X-ray satellites launched that were used by astronomers to make detailed studies of particular objects and particular regions of the sky. They include the Einstein Observatory (original called HEAO-2) launched in 1978, the Chandra Observatory (originally called AXAF) launched by NASA in 1999, and XMM-Newton launched by ESA in 1999. The latter two are still operating successfully as of 2016.

Chapter 10
Gamma Ray Surveys

Gamma-rays are the shortest wavelength and highest energy photons known to physics. A reasonable wavelength boundary between X-rays and gamma-rays is at a photon energy of 100 keV, while a helpful physics distinction is that atomic processes—i.e. interactions between electrons and nuclei—usually produce X-rays, while purely nuclear reactions usually produce gamma rays. The wavelength range of gamma rays observed by astronomers is extremely wide (see Figure 10.1), and data in some parts of the spectrum are scarce. Because a single gamma ray carries such a lot of energy, there are many fewer of them in the universe than there are, say, radio or infrared photons; every gamma-ray photon that a telescope receives is an individually recorded event.

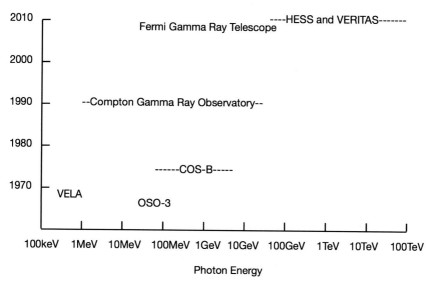

Fig. 10.1 Dates and wavelength coverage of major gamma ray surveys

© Springer International Publishing Switzerland 2016
G. Wynn-Williams, *Surveying the Skies*, Astronomers' Universe,
DOI 10.1007/978-3-319-28510-8_10

Most astronomical gamma rays originate from collisions between a cosmic ray particle—usually a proton traveling very close to the speed of light—with either another particle or with a lower energy photon. Gamma ray astronomy is therefore almost entirely concerned with exotic objects such as neutron stars and active galaxy nuclei, which have the conditions under which protons can be accelerated to these high energies.

10.1 Early Surveys

The first gamma-ray satellite, Explorer 11, was designed by William Kraushaar, and George Clark of MIT and launched in 1961. During its 7-month lifetime it collected 22 gamma rays that could not be accounted for by spurious emission from the Earth's atmosphere. These photons appeared to come from all over the sky rather than one particular direction.

The first gamma-ray map of the whole sky was made by NASA's OSO-3 satellite, built by the same MIT group. It sent data back to Earth from 1967 to 1969. It was sensitive to gamma rays with energies above 50 MeV and detected a total of 821 of them. The resulting map (Figure 10.2) was good enough to demonstrate that gamma rays came from all over the sky, but with an excess corresponding to the plane of the Milky Way galaxy.

The SAS-2 satellite, which was launched in 1972, succeeded in identifying several discrete sources of gamma-ray emission, including the Crab Nebula and the Vela pulsars. It confirmed the concentration of sources towards the galactic plane, but it failed before it could complete a map of the whole sky.

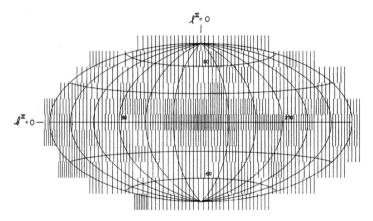

Fig. 10.2 OSO-3 map of the sky in galactic coordinates. The number of vertical bars indicates the strength of the emission in that direction ©American Astronomical Society, from ApJ 177, 341 (1972)

10.2 COS-B

The COS-B satellite was launched by the European Space Agency in 1975 and sent back data for seven years. It was sensitive to gamma rays with energies between 70 MeV and 5 GeV. It detected emission from several pulsars and supernova remnants, and at least one quasar, 3C273. Its most spectacular success, though, was its map of the gamma-ray emission from the plane of the Milky Way galaxy (Figure 10.3). The map bears a strong resemblance to the map of molecular hydrogen (Figure 5.15). To understand why this is so we need to introduce the subject of cosmic rays.

Cosmic rays are charged particles that are traveling through space at almost the speed of light. The word "ray" is a misnomer since cosmic rays are not photons; most of them are protons or electrons. They are blocked from reaching the Earth's surface by our atmosphere, but have little problem traversing the Galaxy. Because they carry electric charge, their paths are wildly deflected by the Galaxy's magnetic fields so that the direction that a cosmic ray is heading when it reaches Earth provides almost no clue as to where it originated. This inability to link a cosmic ray to its place of origin is one reason that direct studies of cosmic rays are usually considered to be a branch of physics rather than astronomy.

The most likely sources for cosmic rays are supernovae in our Galaxy and active nuclei in other galaxies. The strong and rapidly-changing magnetic fields in these objects can accelerate charged particles to very high energies. If a cosmic ray proton of sufficient energy subsequently collides with an interstellar gas nucleus, a nuclear reaction takes place that produces gamma rays. The gamma rays, being photons, are not affected by the Galaxy's magnetic field, so travel in a straight line to us from the location where they were formed. Figure 10.3 is therefore primarily a map of the places where galactic cosmic rays are colliding with interstellar gas atoms—mainly hydrogen and helium. From the radio maps of CO molecules and of neutral hydrogen we can estimate the amount of interstellar gas in any direction, and from the gamma-ray maps we can tell how many collisions are occurring there. By comparing these numbers we can work out the density of cosmic rays all over the Galaxy. What we find is that the flux of cosmic rays in the outer parts of the Galaxy are roughly similar to that in our own neighborhood, but that it increases in the central regions of the Galaxy. One of the implications of this result is that a large fraction of cosmic rays we detect must originate from within our own Galaxy. The most likely sources are supernova explosions.

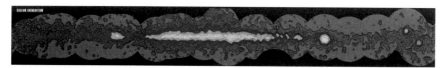

Fig. 10.3 COS-B map of the galactic plane. The brightest point source in the picture is the Vela pulsar. The two fainter point sources on the far right are the Crab and Geminga pulsars. Image credit: European Space Agency

10.3 Compton and Fermi Satellites

The next major gamma-ray survey was that made by the Compton Gamma Ray Observatory. At the time it was launched in 1991 it was the heaviest astronomical satellite ever built, weighing some 17 tons. It contained four instruments, two of which were used to make all-sky surveys. They were the COMPTEL imager which operated in the range 1–30 MeV, and EGRET which operated in the 20 MeV–30 GeV range. It also carried an instrument called BATSE that was specifically designed to study gamma ray bursts; we will discuss it in the next section.

EGRET, which stands for "Energetic Gamma Ray Experiment Telescope" made the first sky survey at energies above 100 MeV. As at the lower energies mapped by COS-B (Figure 10.3), emission from the galactic plane dominates the sky. With its higher sensitivity and resolution, however, EGRET also found 271 compact gamma-ray sources, at least 66 of which have been identified with active galaxy nuclei of a sort called "blazars" (Figure 10.4). Blazars are related to quasars and were originally discovered by radioastronomers as compact variable objects in the centers of certain elliptical galaxies. Matter falling onto a supermassive black hole in the galaxy nucleus releases gravitational energy that is focussed into a pair of narrow jets along the axis of the disk (Figure 10.5). The blazars that we observe in radio or gamma rays are those in which the jet happens to point towards the Earth.

Not all the compact EGRET gamma-ray sources can be identified with active galaxy nuclei. Five of them are pulsars; most of the rest are unidentified but show a concentration towards the galactic plane. They are therefore most probably connected with supernova-related phenomena in our own Galaxy.

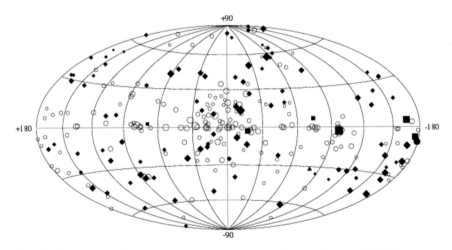

Fig. 10.4 Map of the point sources of gamma-ray emission discovered by the EGRET instrument on the Compton Gamma Ray Observatory. Black squares are pulsars, black diamonds are blazars and open circles are unidentified objects. Image credit: NASA/HEASARC

Fig. 10.5 Artist's impression of a "blazar." Image credit: NASA

The Compton Gamma Ray Observatory operated successfully until 2000. Its successor is the Fermi Gamma Ray Telescope, a NASA satellite, which was launched in 2008 and is still fully operational as of 2016. Its instruments are designed both to survey the sky and to conduct detailed studies of specific objects. The whole-sky map it produced (Figure 10.6) is the most detailed gamma-ray map we possess.

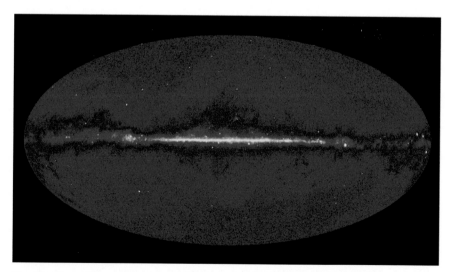

Fig. 10.6 Fermi Gamma-ray Telescope map of the whole sky, showing both the extended galactic emission and the extragalactic point sources. Image credit: NASA/DOE/Fermi LAT Collaboration

10.4 Gamma Ray Bursts

After the ratification of the nuclear test-ban treaty of 1963 the USA proceeded to launch a series of six, and later twelve military satellites designed to detect illicit nuclear explosions. These VELA satellites contained instruments that would detect X-rays, neutrons, light flashes, and 0.2–1.5 MeV gamma rays emitted from atmospheric nuclear explosions anywhere in the world. They orbited the Earth at high altitude—a third of the distance to the Moon—and could determine the location of an explosion by the minute time differences between signals received by different satellites.

On July 2nd 1967 two of the satellites detected a flash of gamma radiation that did not have the expected properties of a nuclear explosion. Scientists at Los Alamos Laboratory were initially not sure what to make of the event, but as more and improved VELA satellites became operational it became clear that these gamma ray bursts did not originate from the Earth, but from distant locations in space. Confident that what they were seeing were astronomical rather than military events, the team decided to declassify the results. Information about sixteen events in the previous four years was published in the astronomical press in 1973.

Most bursts lasted for only a few seconds or less (see Figure 10.7), but during that time they could outshine every other gamma-ray source in the sky. These events were uncorrelated with known supernova explosions during this period, so had to originate from some other kind of cosmic event.

These bursts generated enormous interest among astronomers and within a very few years several satellites were built or adapted to observe the phenomenon. These satellites were able to measure much more accurate positions of bursters in the sky but, frustratingly, none appeared to be associated with any plausible source, such as a pulsar, supernova remnant or active galaxy.

Major progress came with the Compton Gamma Ray Observatory, which we discussed in the previous section. It carried a specially-designed instrument called BATSE, for Burst and Transient Source Experiment, which detected 2700

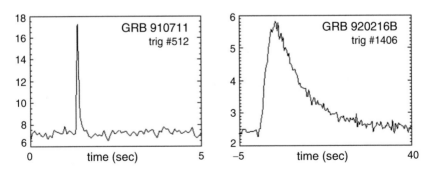

Fig. 10.7 Two examples of gamma-ray bursts, showing a range of durations. Image credit: NASA

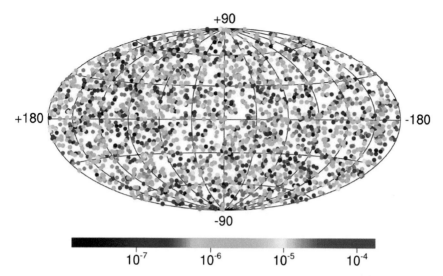

Fig. 10.8 Map of gamma ray bursts detected by the BATSE instrument in the Compton Gamma Ray Telescope. The colors indicate the great range in intensity of the bursts. Image credit: NASA/HEASARC

gamma-ray bursts during its nine-year life. The location of these events was found to be remarkably uniform over the whole sky (Figure 10.8) indicating that they must originate from distances far beyond our own Galaxy, and even beyond nearby clusters of galaxies such as the Virgo Cluster. The alternative possibility, that the bursters were near neighbors of the Sun within our Galaxy, has been excluded by other observations.

Several theories for the origin of these bursts predicted that there should be an afterglow at longer wavelengths for minutes or hours after the burst. With this in mind an Italian-Dutch consortium launched the BEPPOSAX satellite in 1996. It was specifically designed to quickly make follow-up observations of bursts at X-ray wavelengths. Its successor, SWIFT, was launched in 2004 and included both X-ray and optical instruments; it is still operational in 2016. Rapid follow-up with these instruments, as well as ground-based telescopes, confirmed the extragalactic nature of the bursters; indeed one of them, GRB090423, has one of the largest redshifts, 8.2, ever observed.

Gamma-ray bursters have a wide range of properties. Some bursts last only milliseconds while others can last for several minutes. The cause of bursters is still very much under discussion; mergers of neutron stars and black holes are among the mechanisms proposed. The interest in gamma-ray bursters is not entirely astrophysical: if a major gamma-ray burst were to occur in the solar vicinity it could have a devastating effect on the Earth's biosphere; indeed it has been speculated that the Ordovician-Silurian extinction event of 450 million years ago could have been caused by a gamma-ray burst.

10.5 Very High Energy Gamma Rays

At the highest gamma ray energies, above 100 GeV, a fundamental problem looms: as discussed in Appendix A.3 an astronomical object will emit far fewer photons if it radiates gamma rays than if it radiates the same power as, say, radio waves. This problem becomes especially severe at energies above about 100 GeV; even the brightest gamma-ray sources in the sky produce so few photons that the chances of one being caught by something as small as a satellite-borne telescope become very small.

The solution to this problem is to make use of a phenomenon called the Cherenkov Air Shower. When an incoming very high energy gamma ray hits the Earth's upper atmosphere a series of nuclear reactions takes place leading to the production of secondary particles that are traveling extremely close to the speed of light. Since the speed of light in a medium—in this case air—is slightly less than that in a vacuum, these particles can find themselves traveling faster than the local speed of light. When this happens, a shock wave that is analogous to a sonic boom is produced. However, instead of sound waves, these particles produce light waves that are referred to as Cherenkov radiation after Pavel Cherenkov (1904–1990) who discovered the effect in his laboratory in 1934. By this process the energy of a single gamma-ray photon is spread among many visible-wavelength photons that travel in nearly, but not exactly, the same direction as the original photon. The Cherenkov light pulse from a single gamma-ray photon is thus spread into a cone about 1 degree wide, illuminating a ground area of thousands of square meters. A single moderately-sized optical telescope within this area therefore can have an effective collecting area that is about 10,000 times greater than that of any current satellite experiments (Figure 10.9).

The first instrument to successfully detect this kind of radiation was a 10-meter diameter optical telescope at the Whipple Observatory in Arizona (Figure 10.10). Since the aim of this telescope was to collect light from the Cherenkov radiation pulse from the Earth's upper atmosphere it did not need to provide particularly sharp images. In order to reduce costs, therefore, the mirror was constructed of several hundred small optical reflectors mounted in a way that made it look more like a radio telescope than an optical one. The first breakthroughs with this telescope were the detection of the Crab Nebula in 1989 and the blazar Markarian 421 in 1992, demonstrating that very high energy gamma ray astronomy encompasses both galactic and extragalactic objects.

Several other very high energy gamma ray observatories have been constructed since then, most notably the VERITAS instrument (Very Energetic Radiation Imaging Telescope Array System), also on Mount Hopkins, and the HESS array (High Energy Stereoscopic System) built by German astronomers and located in Namibia, southern Africa (Figure 10.11). Both these instruments consist of several multi-mirror telescopes spread over an area hundreds of meters across. The multiple images can be used stereoscopically to obtain a three-dimensional image of the particle shower at the top of the Earth's atmosphere, greatly improving the positional accuracy. The largest reflector in the HESS array has 875 hexagonal

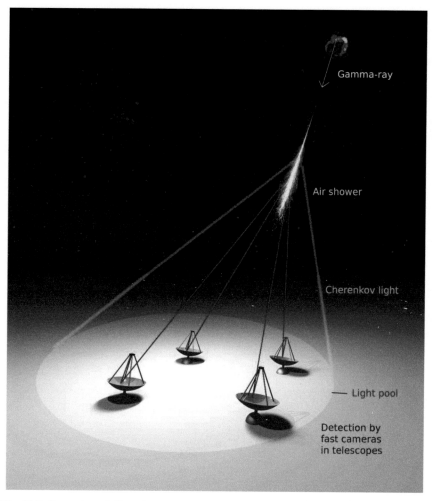

Fig. 10.9 Generation of Cherenkov radiation by a very high energy gamma ray hitting the top of the Earth's atmosphere. The spread of the cone is greatly exaggerated. Image credit: Konrad Bernloehr

mirrors giving an effective diameter of 28 meters—far larger than any conventional optical telescope; however, the images it produces are a couple of minutes of arc across—excellent for its purpose, but nowhere near the quality of an instrument such as the Keck Telescope.

The science of very high energy gamma rays is so new that there has not yet been an unbiased survey of the whole sky. Most of the sources shown in Figure 10.12 were detected as a result of deliberately pointing the array at objects already discovered at other wavelengths. However the HESS instrument did make an unbiased map of a $115° \times 7°$ region around the galactic center; this explains why Figure 10.12 shows so many sources in that region. The majority of the sources in

Fig. 10.10 The Whipple gamma ray telescope on Mount Hopkins in Arizona. Image credit: Stephen Fegan

Fig. 10.11 The HESS array in Namibia. Image credit: HESS Collaboration, Clementina Medina

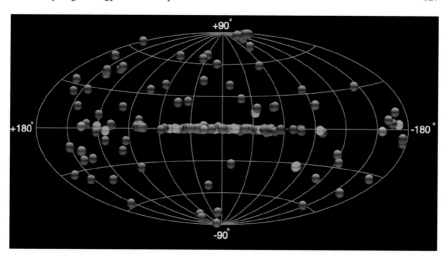

Fig. 10.12 Very high energy gammas ray sources detected by HESS as of mid 2015. The colors refer to different types of object: the most common, in red, are active galaxy nuclei. Image credit: Scott Wakely & Diedre Horan, TeVCat

this galactic plane mini-survey are supernova- or pulsar-related, but about a third are unidentified. The extragalactic objects which were successfully detected are mainly quasars, blazars, and radio galaxies. We do not yet know if we will find any new classes of astronomical object at these wavelengths; the next ten years will be an exciting time for this science.

There is no fundamental reason why there cannot be gamma rays at energies even higher than those detected by VERITAS and HESS. The highest energy gamma ray that has been detected until now has an energy of about 100 TeV, but the highest energy cosmic ray particle (probably a proton) that has been detected by physicists had an energy of some 300 million TeV—about the same as a baseball traveling at 60 mph. There are almost certainly forces in the universe capable of generating photons with energies much greater than anything astronomers have detected so far; how long it will be before we detect them remains to be seen.

Chapter 11
Space Astrometry

Most of the surveys discussed in this book have the goal of discovering new objects in the sky. The surveys in this chapter, however, focus on astrometry—the accurate measurement of star positions. But while astrometry in the telescope era (Chapter 3) was driven by the need to improve navigation, its focus now is the accurate determination of parallaxes and proper motions of already-known stars. From the parallaxes we can derive more accurate distances—and hence luminosities—of the stars. From the proper motions we can study the clustering and dynamics of the stars in our galaxy. We can also search for evidence of planets or binary companions that cause the position of a star to oscillate in space (Figure 11.1).

The enormous improvement in astrometric data over the last 25 years comes from the use of space telescopes that avoid the blurring effects caused by the Earth's atmosphere. Since these telescopes are in space they could in principle make their observations at almost any wavelength. It turns out, however, that the optimum wavelength for space astrometry is the visible region: at longer wavelengths blurring by diffraction becomes a problem, while at shorter wavelengths detector systems are less sensitive, and most stars give out less radiation.

11.1 Hipparcos

The Hipparcos satellite was launched in 1989 by the European Space Agency and orbited the Earth for 3.5 years. The name is an acronym for High Precision Parallax Collecting Satellite as well as a homage to the Greek astronomer Hipparchus (Chapter 2). The optical system in the satellite was unconventional, consisting of a 29-cm diameter telescope that collected light simultaneously from two small patches of sky 58° apart as the satellite slowly rotated around the sky (Figure 11.2). The precise separations of millions of pairs of stars, such as Star A and Star B, was collected in a database that was then used to calculate greatly improved positions of each individual star as it moved across the sky during the time that Hipparcos was in orbit.

© Springer International Publishing Switzerland 2016
G. Wynn-Williams, *Surveying the Skies*, Astronomers' Universe,
DOI 10.1007/978-3-319-28510-8_11

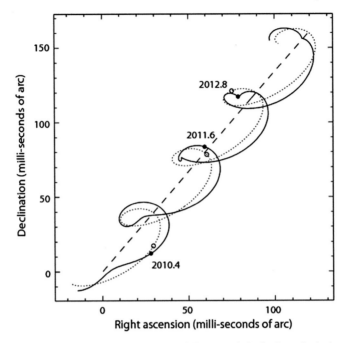

Fig. 11.1 The apparent position in space over a 3.5 year period of a hypothetical star with an exoplanet. The straight dashed line from bottom left to top right represents the system's proper motion. The dotted line includes the effect of parallax. The solid line includes the effects of the gravitation pull on the star from an invisible planetary companion. Image credit: Michael Perryman

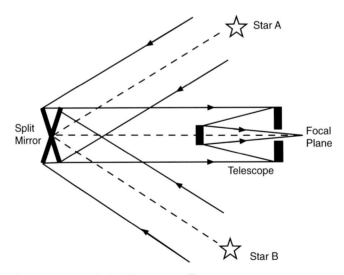

Fig. 11.2 Mirror arrangement in the Hipparcos satellite

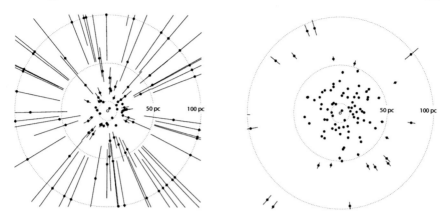

Fig. 11.3 Distances to nearby stars with their error bars based on ground-based observations *(left)* and Hipparcos observations *(right)*. Image credit: Michael Perryman

The main focus of the Hipparcos program was a master list of 118,000 stars spread over the whole sky. The positions of these stars were determined to an accuracy of 1 milliarcsecond, an angle similar to that subtended by an astronaut standing on the Moon's surface as seen from the Earth. About 2.4 million more stars were measured to somewhat lower accuracy, including 99 % of the stars that are brighter than 11th magnitude. Improved values for the proper motions of all these stars were determined by comparing their Hipparcos positions with those measured nearly a century earlier in the Astrographic Catalog of the Carte du Ciel (see section 4.1).

Figure 11.3 gives a good example of the enormous improvement Hipparcos has provided to our knowledge of distances of stars up to 100 parsecs away. The stars featured in Figure 11.3 are all exoplanet host stars. The left panel shows their estimated distances from the Sun based on pre-Hipparcos ground-based parallax measurements, while the right-hand panel shows the distances and error bars based on Hipparcos. The improved distances in the right-hand panel allow for much more reliable estimates of the luminosity and nature of the stars in question, refining the search for systems with Earth-like properties.

Several thousand scientific papers have been written based on the results of the Hipparcos survey. It has improved our understanding of the dynamics of the Galaxy and the structure of its spiral arms. It also provided evidence of a possible galaxy merger in our past.

Hipparcos's scientific legacy even extends to cosmology. The distance to a galaxy is often determined by identifying a certain type of star (the so-called "Cepheid variables") and measuring how much fainter they are in that galaxy than in nearby regions of the Milky Way. By redetermining the distances to the nearby Cepheid variables Hipparcos has allowed us to recalculate the distances to many galaxies in our vicinity—thereby improving our understanding of how galaxies are grouped into clumps and filaments.

11.2 Gaia

The successor to Hipparcos is the European Space Agency's Gaia satellite (Figure 11.4) which was launched in December 2013 and is currently undertaking an all-sky astrometric survey that promises to be a major advance on Hipparcos. Its main scientific goal is to create a three-dimensional map of a large fraction of the Milky Way Galaxy based on observations of about a billion stars, or about 1 % of all the stars in the Galaxy.

Gaia has an orbit similar to the millimeter-wave satellites WMAP and Planck, orbiting the Sun near the second Lagrangian point (Figure 7.7). At this location it stays at a fixed distance from both the Earth and the Sun, which are both in the same direction in the sky; because it is never shadowed from the Sun it can maintain great thermal stability.

If it works as planned Gaia, will determine the positions, parallax distances, and proper motions to much greater accuracy than ever achieved up till now. Positions of stars brighter than 10th magnitude should have positions measured to 7 microarcseconds, 100 times better than Hipparcos. The parallax distances to about 20 million stars will be measured with a distance precision of 1 %, and stars as far away as the galactic center will have their distances measured to 10 %.

Fig. 11.4 Artist's impression of the Gaia spacecraft in orbit. The large disk, which is about 10 m across, is a sunshield. Observations are made through the apertures in the sides of the central section of the satellite. Image credit: ESA, D. Ducros

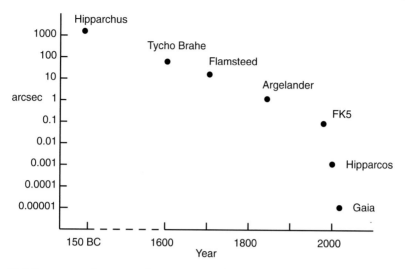

Fig. 11.5 Improvement in astrometric accuracy over time. Based on a graph by Erik Høg

Figure 11.5 shows a graph of how astrometric accuracy has improved over time. It includes the early works of Hipparchus and Tycho Brahe that we discussed in Chapter 2, plus those of Flamsteed and Argelander that we discussed in Chapter 3. The point marked FK5 refers to the final catalog in a long series of ground-based astrometric studies initiated by the German astronomer Arthur Von Auers in the 1860s. The enormous improvements in accuracy made by Hipparcos and Gaia are each greater than any that occurred before the space age.

Chapter 12
The CCD Era

In Chapters 5–10 we described the opening up of the electromagnetic spectrum to astronomers, partly as a result of the ability to put telescopes in space. It is now time to enter the fifth era of astronomy, in which visible light has once again become a cornerstone of progress. and ground-based telescopes are once again taking the lead in conducting astronomy surveys. A key to this resurgence has been the development of the charged coupled device (CCD)—a solid-state chip that is capable of capturing images electronically and then feeding them into a computer where they can be stored and analyzed. CCDs, and related devices called C-MOS chips, have completely replaced photographic plates as the technology of choice for recording visible and near-infrared images and spectra.

CCD devices were invented at the AT&T Bell Laboratory in 1969, and were in routine use on some telescopes by the end of the 1980s. A CCD device consists of a matrix of silicon cells mounted behind a light-sensitive surface which converts incoming photons into electric charges. The CCD cells accumulate the electrons for a chosen length of time before being read out and counted in a computer system. Early devices used in astronomy had areas of only a few hundred pixels across, which made them good for studying specific objects such as individual galaxies, but less useful than photographic plates for mapping large areas of sky. But as CCDs grew in size and shrank in cost, astronomers were able to mount them in arrays that contained tens and later hundreds of millions of pixels; by the year 2000 there were very few professional astronomers still using photographic plates.

Visible-wavelength CCDs have been crucial to the success of a number of astronomy satellites including the Hubble Space Telescope, but the large-scale surveys that are the basis of this chapter are all ground-based. Ground-based telescopes have several advantages over space-based ones when it comes to making large scale surveys. Besides the obvious ones—lower cost, larger size, and repairability—a ground-based telescope has the advantage that its data-handling rate is not limited by the bandwidth of its radio communication system. For the surveys described here, these advantage outweigh the image blurring produced by the Earth's atmosphere and the frustrations of bad weather.

© Springer International Publishing Switzerland 2016
G. Wynn-Williams, *Surveying the Skies*, Astronomers' Universe,
DOI 10.1007/978-3-319-28510-8_12

12.1 Sloan Digital Sky Survey

The first major survey to be carried out using CCD detectors was the Sloan Digital Sky Survey (SDSS) which makes use of a specially-built 2.5-meter diameter telescope installed at Apache Point Observatory in New Mexico (Figure 12.1). The survey was started in 2000 and is continuing.

The main goal of the Sloan Survey is to build a three-dimensional map of the galaxies in the local universe. This is done by making two kinds of observation. A particular piece of sky is first imaged using five different color filters and a digital camera system that has thirty 2048×2048 pixel CCDs. From these images the positions and brightnesses of the galaxies in that part of the sky are catalogued. The camera is then replaced by a spectrograph which can measure the redshifts up to 1,000 galaxies at a time.

Obtaining 1,000 spectra at a time requires some detailed preparation: for each patch of sky to be observed a special metal plate is prepared with a small hole

Fig. 12.1 Sloan Telescope New Mexico. The unusual shape of this telescope is due to its wind shields. When the telescope is not in use it is covered by a movable shed that runs on the tracks leading off the bottom left of the picture. Image credit: Fermilab

Fig. 12.2 A technician connects the fibers between the spectrograph and a metal plate that will be mounted at the focus of the Sloan Telescope. Image credit: New Mexico State University

drilled for each galaxy in the field. The plate is then installed at the focal plane of the telescope, and each hole is connected to an optical fiber that carries the light to a spectrograph that is also mounted on the telescope (Figure 12.2). The spectrograph then measures and records the light from all 1,000 galaxies simultaneously.

The resulting galaxy spectra can be used in several different ways; they can provide information about the types of stars in the galaxy and they can find quasar-like activity in their nuclei. But the single most important piece of information that a galaxy spectrum provides is its redshift, since this can be used to determine the distance to the galaxy using Hubble's Law.

Figure 12.3 shows the distribution of galaxies that lie within 1.25° of the celestial equator and within a distance, determined from their redshift, of about 6,000 megaparsecs. The blank areas to the left and right correspond to the directions where the Milky Way crosses the equator; the SDSS generally ignores these regions, since extinction by interstellar dust in these directions greatly inhibits the study of external galaxies. The data for the rest of the sky show that galaxies are very strongly grouped in clusters and strings and sheets. These patterns are of great interest to cosmologists, who are using them to study problems such as the distribution of dark matter in the universe.

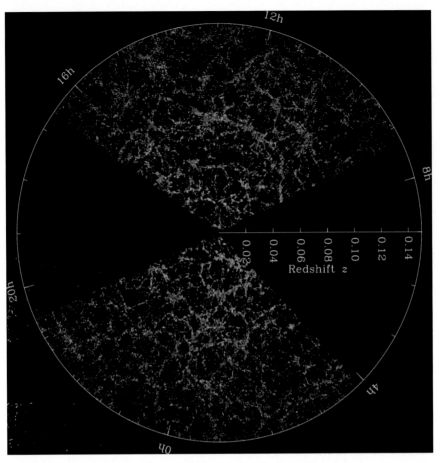

Fig. 12.3 Sloan Digital Sky Survey map of the sky showing the distribution of galaxies in a thin slice between declinations 1.25°N and 1.25°S. A redshift of 0.10 corresponds to a distance of about 400 Mpc. Image credit: M. Blanton and the Sloan Digital Sky Survey

As of July 2013 the SDSS had surveyed about one third of the whole sky, measuring the position, brightness, and color of over 200 million galaxies down to a limit of around 22nd magnitude. It had obtained the spectra of over 800,000 galaxies and 100,000 quasars and made them available online to anyone who is interested. It is continuing to gather spectra from northern hemisphere galaxies as well as certain parts of the Milky Way Galaxy. It is also expanding its reach into the southern hemisphere by taking over responsibility for the existing 2.5-meter Irénée du Pont telescope in Chile and installing new instruments on it.

12.2 Pan-STARRS

Another major sky survey that exploits the advantages of CCDs is the University of Hawaii's Pan-STARRS, which stands for Panoramic Survey Telescope and Rapid Response System. It is situated on the summit of Mount Haleakala on the island of Maui (Figure 12.4).

The Pan-STARRS survey, which began in 2010, takes a different approach to that of SDSS in several major ways. Firstly it maps all the sky, both galactic and extragalactic, that is observable from Hawaii, and secondly it makes regularly repeated observations of the sky, making it an extremely powerful tool for discovering and studying moving objects such as asteroids, and time-varying objects such as variable stars, supernovae, and quasars.

The Pan-STARRS telescope has a 1.8-meter diameter mirror that focuses light onto what is currently the world's largest digital camera. The focal plane of the camera (Figure 12.5) comprises 60 silicon chips, each of which contains 64 CCDs, giving a total of approximately 1.4 billion pixels. The pixels cover an area 40 centimeters square, which is significantly larger than the area of Palomar Sky Survey image (Figure 4.6). In comparison, a typical domestic digital camera contains something like 5 million pixels on a chip a few millimeters across.

A typical exposure of the Pan-STARRS camera takes 30 seconds and produces 2 gigabytes of data. The images are used in three ways: first they are added to previous images taken of that piece of sky to build up a very sensitive long-exposure

Fig. 12.4 The Pan-STARRS telescope on Mount Haleakala in Hawaii. The mountain in the distance is Mauna Kea. Image credit: Robert Ratkowski and PS1 Scientific Consortium

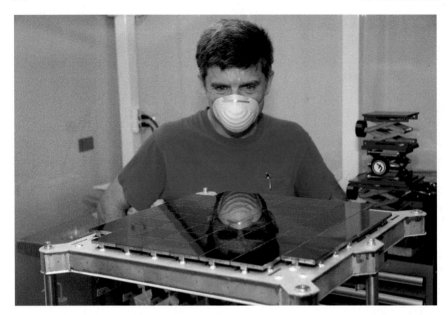

Fig. 12.5 The focal plane array of CCDs before it was installed in the camera of the Pan-STARRS telescope. Image credit: John Tonry, University of Hawaii

image. Secondly, each new image is compared to recent ones and scanned for differences. Thirdly, all images are stored in a publicly-available database. Most of the sky has been imaged 60–80 times since 2010, though extra time was spent on several patches of sky of particular interest, such as the Andromeda galaxy.

Figure 12.6 is an example of a cumulative image generated from many separate exposures using several color filters. Although small patches of sky have been mapped in more detail—by the Hubble Space Telescope, for example—the Pan-STARRS cumulative images constitute by far the most detailed map of the whole northern sky, as well as all of the southern sky north of $-30°$ declination.

The greatest strength of the Pan-STARRS survey is its unrivaled ability to spot changes in the sky. The changes may be due to stars or quasars varying their brightness, to supernova explosions, to extrasolar planets transiting in front of their parent star, or to asteroids and comets moving in their orbits around the Sun. Indeed, a major justification for the construction of Pan-STARRS was its ability to detect asteroids that are potentially hazardous to the Earth, a topic which we will discuss further in the next section.

Between 2010 and 2014, Pan-STARRS was managed by a consortium of sixteen universities and institutes from the USA, UK, Germany, Hungary and Taiwan. They embarked on a wide-ranging series of projects using the growing Pan-STARRS dataset, the results of which are just beginning to appear in the scientific literature. During 2016 the whole science database was transferred to the Space Telescope Science Institute along with a complete collection of images, and made available to

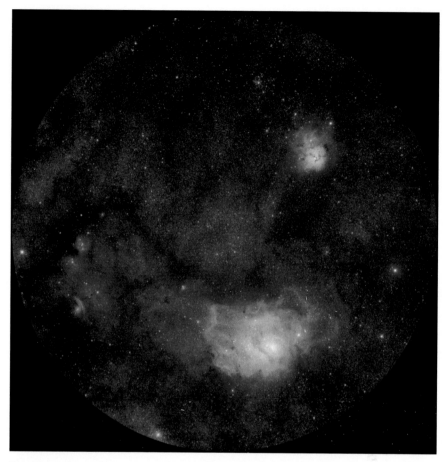

Fig. 12.6 A single field from the Pan-STARRS survey. The image is approximately 3° across—about six times the diameter of the full moon. The visible nebulae include the Lagoon Nebula (Messier 8) and the Trifid Nebula (Messier 20). Image credit: PS1 Scientific Consortium

anyone willing to learn how to use it (see Chapter 14). The total size of the dataset is about 2000 terabytes—about 40 times that of the Sloan survey. It includes images with details as small as one second of arc across, and photometry and variability information in five colors of about 3 billion stars and galaxies that are brighter than 22nd magnitude.

In 2015 Pan-STARRS, now augmented by an almost identical second telescope, embarked on a new NASA-funded project, spending 90 percent of its time searching for near-Earth asteroids, a subject whose recent history we take up in the next section.

12.3 Asteroid Searches

Although the first asteroids were discovered in the early 19th century (section 3.1) it was not until well into the 20th century that the possibility of catastrophic Earth-asteroid collisions was seriously considered. The idea was supported by detailed geological studies of sites such as Tunguska in Siberia and Barringer Meteor Crater in Arizona (Figure 12.7), and became a topic of widespread interest following Louis Alvarez's 1980 proposal that the dinosaurs were rendered extinct as a result of the impact of a large asteroid 66,000,000 years ago. Table 12.1 lists several well-studied events, together with estimates of the sizes of the incoming objects. The Tunguska event, which occurred in Russia, may have been the result of a comet exploding in the atmosphere rather than an asteroid hitting the ground.

The great majority of asteroids orbit the Sun in the asteroid belt between Mars and Jupiter (Figure 12.8). However a significant minority have elliptical orbits that approach or cross the Earth's orbit. Astronomers define Near-Earth-Objects (NEOs) as those asteroids and comets whose orbits bring them within 1.3 astronomical units to the Sun. A subset of these are called Potentially Hazardous Asteroids (PHAs); these are objects which are larger than about 150 meters across and whose calculated orbits will at some point bring them within about 7,000,000 km of the Earth's orbit. There are currently over 1500 known PHAs.

The first CCD-based asteroid survey was the University of Arizona's Space-watch, which started its observing program in 1983. It initially used an 0.9-meter

Fig. 12.7 The Barringer Meteor Crater in Arizona. The crater is 1.2 km across. Image credit: Shane Torgerson

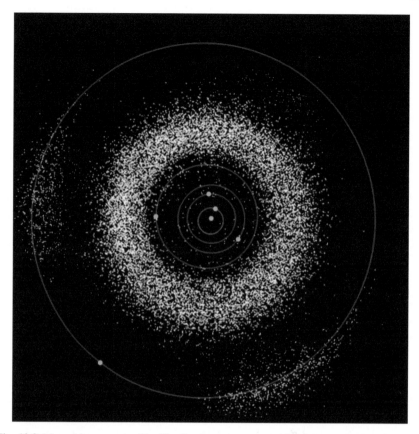

Fig. 12.8 Map of known asteroids. The five large spots near the center are the Sun, Mercury, Venus, Earth, and Mars. The large spot at the lower left is Jupiter. Image credit: Wikipedia

telescope at Kitt Peak Observatory in Arizona, adding a 1.8-meter telescope in 2000. Specially-written software was written to scan images of the sky taken about 30 minutes apart and pick out objects which had moved in the sky during that time. By carefully measuring the position of the asteroid over a number of days its orbit could be calculated. Spacewatch discovered a significant number of both NEOs and PHAs, but the NEO discovery rate increased sharply with the opening in 2000 of LINEAR—a joint project of MIT's Lincoln Laboratory and the US Air Force. The acronym stands for Lincoln Near-Earth Asteroid Research. LINEAR used three 1-meter class telescopes located at the White Sands Missile range in New Mexico and discovered more than 2000 NEOs.

Five years later LINEAR was itself overtaken by University of Arizona's Catalina Sky Survey, which holds the current record for the most NEOs discovered. Catalina uses two telescopes in Arizona plus one in Australia. In 2008 it discovered an object that was on a direct path to the Earth, and made the first successful prediction of an imminent asteroid impact. Fortunately, this one—named 2008 TC$_3$—was not dangerous; it had a diameter of about 4 meters and landed in an uninhabited part of the Sudanese desert in Africa.

The most powerful asteroid survey telescope is currently Pan-STARRS, which was described in the previous section. Its strength is in looking for distant NEOs that could be a threat to the Earth years or decades in the future. It spends most of its time searching for moving objects within 20° of the ecliptic plane.

An alternative approach to asteroid surveying is taken by the University of Hawaii's new ATLAS project. It is specifically designed to look for small asteroids that are on an imminent collision course with the Earth, but are too small (a few tens of meters across) to have been detected by Pan-STARRS when they were at a large distance from the Earth. Its unique feature is that it scans the whole sky visible from Hawaii several times every night, weather and lunar phase permitting, thereby picking up fast-approaching objects which might be missed by the surveys with a slower cycle time. Its two 50-cm diameter telescopes are situated about 140 km apart, opening the possibility of parallax distance measurements of asteroids which are sufficiently close to the Earth. ATLAS should provide us with 3–9 days warning for an impact by a 50-m diameter asteroid and 10–40 days warning for a 140-m diameter one.

The cameras on the Catalina, Pan-STARRS, and ATLAS telescopes are sensitive enough to detect thousands of harmless as well as potentially dangerous asteroids each night. Their associated computer systems automatically measure the paths of these objects in the sky and try to identify them with the orbits of the more than 700,000 asteroids whose orbits are already known. The astrometric data on any previously-unknown asteroids that they find are quickly passed to the Minor Planet Center (MPC), located in Cambridge, Massachussets. If the MPC decides that the calculated orbit of the newly discovered object poses a potential threat to the Earth a message is sent to the Jet Propulsion Laboratory in Pasadena, California, which sends out urgent messages to other observatories asking them to observe the object, and issuing warnings if necessary. This arrangement was successfully put to the test by 2008 TC$_3$, the asteroid which landed in the Sudanese desert in 2008. During the 19 hours between its discovery by Catalina and its impact on the ground, its path was observed over 500 times by 27 amateur and professional astronomers around the world.

Event	Date	Estimated Size
Dinosaur extinction	66,000,000 BCE	10 km
Baringer Meteor Crater	50,000 BCE	50 m
Tunguska	1908 AD	50 m
Chelyabinsk	2013 AD	17 m
2008 TC$_3$	2008 AD	4 m

Table 12.1 Selected historic asteroid impacts

12.4 LSST

The Large Synoptic Survey Telescope (LSST) currently under construction in Chile, will be the successor to Pan-STARRS as the world's most powerful visible-light survey instrument (Figure 12.9). When completed it will have an 8.4-meter diameter mirror and a camera with 3.2 billion pixels.

The scientific goals of the LSST project are very broad, including solar system studies, galactic structure studies, time variability studies, and cosmology. The project is overseen by a consortium of more than 35 universities, research institutions, corporations, and individuals, and is headquartered in Tucson, Arizona. It is scheduled to go into full operation in 2022.

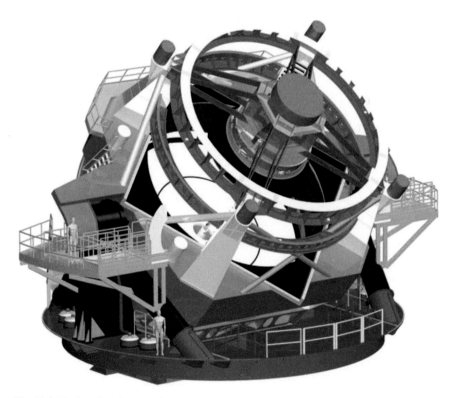

Fig. 12.9 Engineering drawing of the LSST. Image credit: LSST Project Office

Chapter 13
Microsurveys

Most of the surveys described in this book cover either the whole sky, or at least a significant fraction of it. Sometimes, however, astronomers can learn a great deal about a particular phenomenon by focussing their attention on a small patch of sky, so long as it contains a large enough number of relevant objects to make it representative.

In this chapter we discuss four important microsurveys that have been performed in the last 30 years. The first three have connections to cosmology, while the fourth concerns the search for exoplanets.

13.1 Hubble Deep Fields

The Hubble Deep Fields are small patches of sky that have been studied by making very long exposures using the Hubble Space Telescope (HST). The goal of these surveys is the study of extremely distant galaxies whose light has taken so long to reach us that they provide us with clues as to what the universe was like when it was only a small fraction of its present age.

The Hubble Space Telescope (Figure 13.1) was launched by a NASA Space Shuttle in 1990 and remains in full use as of 2015. For the first three years of its life its performance was severely affected by spherical aberration in its optics, but following a successful servicing flight in 1993 it has provided the world with the some of the sharpest and deepest sky images ever obtained. Since the light reaching the telescope is not blurred by the Earth's atmosphere the images are diffraction-limited and can resolve details as small as 0.06 arcseconds across.

The superb image quality of the HST gives it two distinct advantages. Firstly, its famous images of galaxies and nebulae have far finer detail that we can see from using a telescope on the ground. Secondly, there is much more chance of detecting a very faint object if all its light can be concentrated into a small number of pixels rather than spread over many.

© Springer International Publishing Switzerland 2016
G. Wynn-Williams, *Surveying the Skies*, Astronomers' Universe,
DOI 10.1007/978-3-319-28510-8_13

Fig. 13.1 The Hubble Space telescope in orbit. Image credit: NASA

Following its servicing mission in 1993 the HST carried two spectrographs and two digital cameras. The spectrographs were used to study already-known stars and galaxies and operated primarily at ultraviolet wavelengths that are obscured by the Earth's atmosphere. The two cameras were mainly used to make high-resolution images of known nebulae and galaxies, but in 1995 a decision was taken to use both cameras together to make a deep multi-color survey of a small and apparently blank patch of sky to look for objects fainter than had ever been detected before.

Several factors went into the choice of where to point the telescope:

- It had to have an almost empty appearance in the best ground-based images
- It had to be well away from the galactic plane, to avoid the effect of interstellar dust
- It had to be far enough north that it could be seen by HST from all points in its orbit.

The chosen patch of sky, which was only 2.5 arcminutes across and situated in the Ursa Major constellation, was then observed continuously for 10 days, using four different color filters between 0.3 and 0.8 μm—i.e. from near ultraviolet to near-infrared. The resulting image (Figure 13.2) includes about 3000 distant galaxies, but less than 20 stars from the Milky Way Galaxy. The faintest objects in this image are about 30th magnitude, or 10 billion times fainter than one can see with the naked eye.

Fig. 13.2 The original Hubble Deep Field observed in 1995. The step-like shape of the image is the result of obscuration caused by the additional mirrors that were needed to compensate for the spherical aberrations in the telescope's primary mirror. Image credit: NASA

Follow-up spectral observations using large ground-based telescopes have revealed that some of these very faint objects have redshifts as large as 5, indicating that the light we are seeing was emitted when the universe was only a tenth of its current age (see Appendix A.10). By comparing the properties of galaxies at different redshifts, and hence at different ages, astronomers can learn about how the average population of the universe has evolved over its lifetime. One result that has come out of this study is that galaxies with high redshifts tend to be smaller and more numerous than present day galaxies, and to contain a higher ratio of gas to stars. These results suggest that many of our present day galaxies, including the Milky Way, formed as a result of the agglomeration of a number of smaller galaxies. More than 900 scientific papers have been published based on this Hubble Deep Field image and on follow-up observations made at radio, X-ray, infrared, and visible wavelengths.

In 1998, a second Hubble Deep Field survey was performed on a similar-sized patch of sky in the southern hemisphere. It showed a very similar statistical distribution of galaxy properties, providing observational support for one of the basic principles of cosmology, namely that the universe has the same properties in whatever direction we look. It also enabled astronomers using southern-hemisphere telescopes, such as those in Chile, to make follow up observations.

The instruments on the Hubble telescope have since been upgraded several times by Space Shuttle astronauts, leading to significant improvements in its sensitivity and wavelength coverage. Following the installation of the Advanced Camera for Surveys (ACS) in 2003, NASA devoted 400 orbits to the Hubble Ultra Deep Field survey (HUDF). This survey went considerably fainter than the original Hubble Deep Field, sampling the universe as it was about 0.5 billion years after the Big Bang and soon after the first galaxies are believed to have been born (Figure 13.3).

On the final Space Shuttle mission to the Hubble telescope in 2009 NASA installed the Wide Field Camera 3 (WFC3) and used it to make an even more sensitive study of the central part of the HUDF field. This survey became known as the Hubble Extreme Deep Field project (HXDF) which was published in 2012. The HXDF is a combination of nearly 3,000 exposures at nine different wavelengths between 0.4 to 1.7 μm—well into the infrared waveband. Some of the exposures had been taken in connection with other observing programs, but many were made especially for this project and are still under intense study by astronomers.

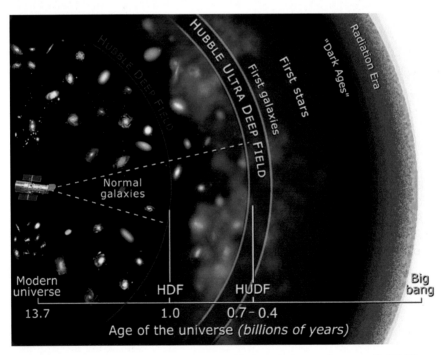

Fig. 13.3 Studying the early universe with the Hubble Space Telescope. Image credit: NASA

   The inclusion of infrared data gave the HXDF a crucial advantage of earlier deep surveys. Most stars, and hence most galaxies, emit the bulk of their power in the near-UV to near-IR wavelength range—say from 0.1 to 2 μm. If a galaxy has a redshift of 6, its energy will arrive at the Earth with a wavelength range of 0.7 to 14 μm—entirely in the infrared waveband.

   Unfortunately, the HST becomes increasingly less useful as we move to longer infrared wavelengths because the effects of diffraction blur its images (see Appendix A.4). The only way to minimize this problem is to build an orbiting telescope with a much larger diameter mirror than the HST. This, indeed, is the main justification for NASA's James Webb Space Telescope (JWST) which is currently under construction (Figure 13.4). The JWST will have a 6.5-meter diameter mirror and will operate at wavelengths between 0.6 μm and 28 μm. One of its main scientific goals will be to push the science of ultra-deep cosmological surveying to even fainter levels than have been achieved by the Hubble Space Telescope. It is scheduled for launch in 2018.

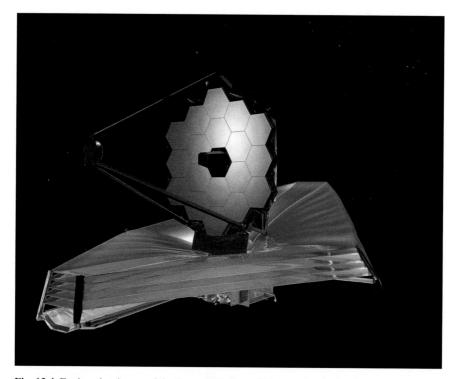

Fig. 13.4 Engineering image of the James Web Space Telescope, currently under final construction. The main mirror, made of gold-plated beryllium, will be 6.5 meters across and will always be shielded from the Sun by a series of five flexible reflective membranes seen beneath it. Both the main mirror and the sunshields will unfold themselves once their launch vehicle has placed the telescope in its proper orbit. Image credit: NASA

13.2 The COSMOS Survey

The Cosmic Evolution Survey, usually referred to as COSMOS, is a program aimed at understanding how galaxies and galaxy clusters have evolved over the last 12 billion years or so. Its goal is broader than that of the Hubble Deep Field surveys discussed in the previous section, which were concerned only with the very earliest epoch of the universe. It is also the only survey in this book to have produced a map of dark matter in the universe, albeit only in a small fraction of it.

The COSMOS survey is a study of a 1.4° square patch of sky near the celestial equator. It is in a region where there are very few Milky Way stars and no obscuring dust clouds. It also has the advantage of being observable by ground-based telescopes in both the northern and southern hemispheres. The first telescope to complete a survey of this region was the Hubble Space Telescope, which used 640 orbits to cover the whole field (Figure 13.5)—the largest contiguous single area of sky it has ever mapped. It detected over 400,000 galaxies.

Follow-up studies have been made by a team of more than 100 astronomers from all over the world using space-borne and ground-based telescopes that cover the complete range of wavelengths from radio to X-rays. The project has been awarded observing time on the following telescopes:

- **Visible and near-infrared wavelength telescopes:** The Keck Observatory, the Subaru Telescope, the Very Large Telescope (VLT), the Canada-France-Hawaii Telescope (CFHT), the University of Hawaii 2.2 m telescope, the United Kingdom Infrared Telescope (UKIRT), and the Cerro-Tololo Inter-American Observatory (CTIO).
- **Far-infrared and submillimeter telescopes:** The Herschel Space Observatory, the Spitzer Space Telescope, the James Clerk Maxwell Telescope (JCMT), the Caltech Submillimeter Observatory (CSO), the Submillimeter Array (SMA), the Institut de Radioastronomie Millimétrique telescope (IRAM), and the Atacama Large Millimeter Array (ALMA).
- **Ultraviolet telescopes:** The Galaxy Evolution Explorer (GALEX).
- **X-ray telescopes:** The Chandra X-ray Observatory, and XMM-Newton.
- **Radio telescopes:** The Very Large Array (VLA), and the Sunyaev-Zeldovich Array (SZA).

The visible and near-infrared telescopes were the major tools for making maps of the distributions of galaxies and measuring their redshifts, and thus their ages. The far-infrared, X-ray, and radio telescopes were used to determine which galaxies contained regions of active star formation, and which contained quasars or black holes in their centers. The X-ray observations were also used to map out the large clouds of very hot intergalactic gas that are often found associated with clusters of galaxies.

One of the major results to emerge from the COSMOS survey is the role of "dark matter" in the evolution of galaxies and clusters of galaxies. The existence of dark matter was first suggested by Caltech astronomer Fritz Zwicky in the 1930s. By comparing the individual Doppler shifts of several galaxies in a cluster he concluded

Fig. 13.5 The COSMOS field as observed by the Hubble Space telescope at a wavelength of 0.8μm in 2003–2004, with the size of the full moon for comparison. On this scale the Hubble Deep Field image (Figure 13.2) would be about 3 millimeters across. Image credit: NASA, ESA and Z. Levay (STScI)

that they were moving around much too fast for the cluster to stay together, unless there was more gravitational matter in the cluster than could be accounted for just by its stars. More evidence for dark matter came in the 1970s when Vera Rubin of the Carnegie Institution of Washington noticed that the stars at the edges of some spiral galaxies were orbiting faster than expected. By the 1990s the evidence for dark matter was overwhelming, but many of its properties were still mysterious What had become clear was that dark matter behaves just like regular matter as far as gravity was concerned, but it otherwise does not interact with atoms or with any other particles. Dark matter is now recognized as an extremely important constituent of the universe. In fact, calculations based on observations of the anisotropy of the cosmic background radiation (Chapter 7) imply that dark matter is about five times as common as ordinary matter in the universe. We don't notice it in our daily lives

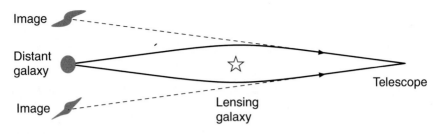

Image

Distant
galaxy

Telescope

Image

Lensing
galaxy

Fig. 13.6 Gravitational lensing of a distant galaxy by an intervening one

because it stays rather smoothly spread out and avoids getting concentrated into
compact objects such as stars, planets, and humans.

Dark matter's inability to create or destroy photons makes it very hard to
study. The COSMOS survey addressed this problem by using the principle of
gravitational lensing. Gravitational lensing occurs when light traveling towards us
from a very distant galaxy is deflected by the gravitational field of an intervening
object. Calculations of the precise path of the light requires use of Einstein's General
Theory of Relativity, but the idea can be understood in principle by considering
Figure 13.6.

As the light from a distant galaxy travels past a massive intervening one, its
path is bent inwards by gravity. It is as though the intervening galaxy is acting like
a converging lens. The light from the distant galaxy will therefore appear to an
observer at the Earth to be emanating from an object to the side of the intervening
galaxy. If the alignment is perfect, and the intervening galaxy is symmetric, the
observer would see a ring round the intervening galaxy, but if the alignment is not
perfect the distant galaxy will appear as one or more distorted images, often in the
shape of an arc. Figure 13.7 shows a particularly clear example of gravitational
lensing, The brightest, roughly circular objects in the image are elliptical galaxies
in the comparatively nearby massive cluster Abell 2218 at a redshift of 0.16, but
most of the fainter streaky features are gravitationally distorted images of galaxies
that are much farther away. Specifically, the bright orange arc near the center is a
distorted image of an elliptical galaxy at a redshift of 0.7, while most of the other
faint arcs are star-forming galaxies at redshifts of 1.0–2.5.

If astronomers know the redshifts of both the distant and the nearby galaxies, they
can use the apparent shapes of the lensed galaxies and the theory of gravitational
lensing to calculate the mass of the intervening material. If the foreground object is
a cluster of galaxies rather than a single galaxy, it is sometimes possible to make a
map of the intervening matter, regardless of whether it is composed of conventional
matter or dark matter; making maps of dark matter in this way (Figure 13.8) has
been one of the most important results to come out of the COSMOS survey.

Fig. 13.7 Gravitational lensing of distant galaxies by a foreground galaxy cluster. Image credit NASA/ESA

By sorting their data by redshift the COSMOS astronomers also established that dark matter has become more concentrated as the universe has expanded. These patterns of the distribution of dark matter are being used by theoretical astronomers to help understand why galaxies in the present universe are clustered into filaments and voids, of the kind seen in the map of the local universe produced by the Sloan Digital Sky Survey (Figure 12.3).

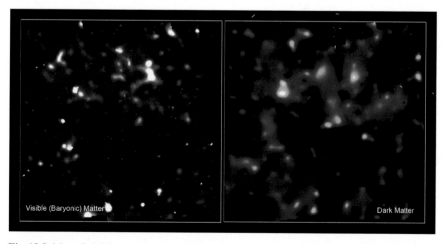

Fig. 13.8 Map of visible matter (left) and dark matter (right) in the direction of the COSMOS field. Image credit: NASA, ESA and R. Massey (Caltech)

13.3 The MACHO Survey

The distributions of visible matter and dark matter in our universe are very different. Dark matter is affected only by gravity, but ordinary matter can be moved by other forces such as gas pressure and radiation pressure. This difference leads to the present situation that dark matter is much more evenly spread out in the universe than ordinary matter.

Surveys such as the COSMOS survey can produce maps of dark matter, but they provide no information about what dark matter is actually made of. As the astronomical evidence for dark matter became stronger in the 1980s and 1990s two main theories arose to try to explain it. According to these two theories dark matter is made up either of "WIMPS" or "MACHOS". The word WIMPS stands for Weakly Interacting Massive Particles, while the word MACHOS stands for Massive Compact Halo Objects.

According to the WIMP theory, the missing mass exists in the form of microscopic fundamental particles that carry no electric charge and do not interact with either protons or photons. This theory implies the existence of one or more completely new subatomic particle and requires a significant modification to some of the fundamental ideas of physics.

In the MACHO theory, on the other hand, the missing mass takes the form of large, discrete objects. They could be black holes or neutron stars, or, more likely, they could be objects that are the sizes of planets or very small, dim stars. Under normal circumstances a star forms when an interstellar cloud collapses under gravity and gets heated by the potential energy that is thereby released. If the cloud is large enough the central temperature can rise to the point where thermonuclear reactions start: the cloud then becomes a star. But if the mass is below about 10 % of the mass of the Sun the pressure and temperature in its center is insufficient for nuclear reactions to take place, and the object remains dark and barely visible. The MACHO theory postulates that the Milky Way and other galaxies might be filled with vast numbers of these dark planet-like objects which give off negligible amounts of light, but which must be significantly more abundant than normal stars. The reference to "halo" in the name MACHO comes from the fact that a number of experiments had already indicated that the dark matter in our Galaxy is more spread out than the starlight we see in the disk and the spiral arms of the Milky Way, occupying roughly the same region of space, often referred to as the "Halo," that globular clusters are found.

Although a MACHO gives out only a negligible amount of light itself, it can have a significant effect on the light from a star that lies behind it. This phenomenon is called gravitational microlensing and is a smaller-scale example of the phenomenon illustrated in Figure 13.6. In this case the distant object is an ordinary star and the lensing object is a MACHO that is too faint to be detected. The gravitational deflections produced by a MACHO are too small to produce visible arcs, but they can produce a significant brightening of the distant star as light from it is focussed towards the observer.

In 1986 the Polish astronomer Bohan Paczynski pointed out that MACHOS in the halo of the Milky Way galaxy should be moving in slow orbits around the galactic nucleus. These motions could occasionally cause them to transit in front of a star in a more distant galaxy. He calculated that these microlensing events would typically last a few days and could lead to a temporary brightening of the distant star by several magnitudes. These events could be distinguished from outbursts such as novae by the fact that the light curves have a characteristic shape, and the colors of the star do not change at all during the event. By counting the frequency of these microlensing events one could calculate whether there were enough MACHOs in the Galaxy to account for the estimated amount of dark matter there.

Three groups set up searches for MACHOS in the early 1990s. In Australia, the so-called MACHO collaboration took over an unused 50-inch telescope at Mount Stromlo observatory and fitted it with a large digital camera. In Chile, a largely French collaboration called EROS, (short for Expérience pour la Recherche d'Objets Sombres) used several medium-sized telescopes in Chile, and a Polish group set up OGLE (Optical Gravitational Lensing Experiment) which also used a telescope in Chile.

The reason that all of these groups were based in the southern hemisphere was their choice of the Large Magellanic Cloud (LMC) as the main source of the background stars in front of which they hoped to see machos transiting. The LMC (Figure 13.9) was ideal for this experiment because it has millions of stars that are bright enough to be individually monitored, and light from the LMC has to traverse a long path through the halo of the Milky Way. The brightness of each of these LMC stars was measured regularly—every night in some cases—with the data stored in

Fig. 13.9 The Large Magellanic Cloud. Image credit: Axel Mellinger, Central Michigan University

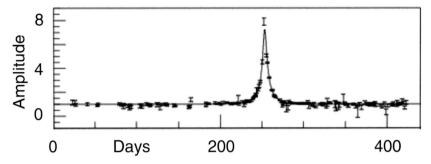

Fig. 13.10 Microlensing event observed by the MACHO collaboration. Image adapted from work by Charles Alcock and colleagues

computers. Searches were then made for stars which brightened and then faded according to the pattern predicted by microlensing theory.

The surveys were a success in that all three groups witnessed several micro-lensing events of the type predicted. Figure 13.10 shows one of these. As the invisible MACHO drifted in front of one of the background stars its brightness rose by a factor of about eight as measured both at red and blue wavelengths. The whole event lasted about a month. About a dozen such events were observed in two years. These events thoroughly vindicated the theoretical predictions of the microlensing effect and showed that there were, indeed, some MACHOS populating the galactic halo. However when the statistics were examined it was clear there were far fewer microlensing events observed than would be predicted if all the dark matter in the Milky Way Halo was in the form of MACHOS. The final conclusion of the surveys was that almost all of the dark matter in the Galaxy's halo is probably made up of WIMPs.

Determining the nature of these WIMPS, however is a major problem in physics, since it impinges on both quantum mechanics, which deals with electrical forces and small-scale phenomena, and general relativity, which is concerned with gravitation and large-scale phenomena. The reconciliation of these two grand theories into a unified picture of the laws of physics will probably have to precede any sophisticated understanding of the nature of dark matter

13.4 The Kepler Exoplanet Survey

The first three "microsurveys" in this chapter are all in some way connected with cosmology—the study of the universe as a whole. This final section brings us much closer to home since its goal is to discover "exoplanets"—planets in orbit around stars other than the Sun.

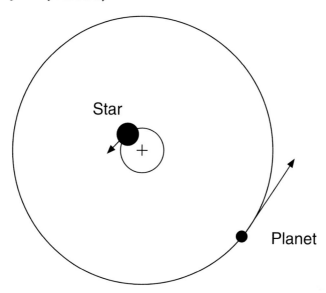

Fig. 13.11 A star and its planet both move in orbits around their common center of mass, which is shown here as a plus sign

For several millennia the Sun was the only star known with confidence to have a planetary system. This situation changed in 1992 when Aleksander Wolszczan and Dale Frail analyzed irregularities in the arrival times of radio signals from the pulsar PSR 1257+12, and showed that these could best be explained if there was a pair of orbiting planets tugging at the central neutron star.

Three years later Michel Mayor and Didier Queloz at the Geneva Observatory announced the first discovery of a planet orbiting a main sequence object, namely the G-star 51 Pegasi. The discovery was made by using a high-resolution visible-light spectrometer to look for subtle changes in the Doppler shift of the star's spectral lines as the planet orbited it; the planet itself is too faint to be seen directly. These changing Doppler shifts occur because in any planet-star system, both objects are moving around their common center of mass; although velocity changes of the star are much smaller than those of the planet, they can be large enough to be detectable (see Figure 13.11). This approach is known as the "radial-velocity technique" and has led to the discovery of over 500 exoplanets.

Taken by itself, however, the radial-velocity technique has a serious limitation. Unless the plane of the exoplanet's orbit lines up exactly with the direction our telescope is pointed, what we measure is an under-estimate of the actual orbital velocity, and any calculations we make of the properties of the planet will be highly uncertain. Fortunately, there is a solution to this problem, which is to specifically search for exoplanet systems in which the plane of the planet's orbit crosses our line

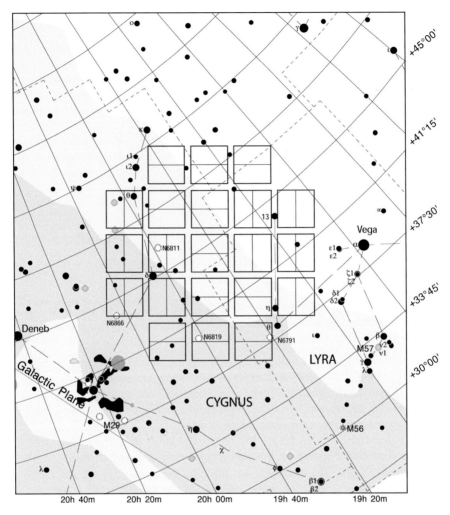

Fig. 13.12 The region of sky observed by the Kepler spacecraft. Image credit: Wikipedia and Software Bisque

of sight to the star. In these cases we can be confident that the Doppler velocity that we measure is equal to the orbital velocity of the star around the system's center of mass. We can find these systems by using the "transit technique," in which a wide-field camera simultaneously monitors the brightness of a large number of stars, looking for the slight dimming that occurs when an otherwise undetected exoplanet crosses in front of its parent star. Stars which show evidence for exoplanet transits are then earmarked for follow-up studies of their radial velocity.

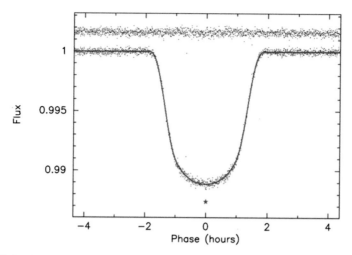

Fig. 13.13 Kepler observations of the transit of a planet in front of a star. Image credit: Michael Endl, ©AAS, Reproduced with permission

While there have been some successful discoveries of exoplanets using the transit technique with a ground-based telescope, most transit discoveries to date have been made by NASA's Kepler spacecraft which was launched in 2009 and is still operating as of 2016. Kepler contains a 1.4-meter diameter telescope with a 95 megapixel camera that continuously monitors a 15 square degree patch of sky situated between the stars Deneb and Vega (Figure 13.12). Kepler is programmed to monitor the brightness of over 145,000 main-sequence stars within this area, most of which have visual magnitudes in the range 14–16. It is looking for the slight dimming of a star that occurs when one of its planets crosses directly in front of a star and blocks a fraction of its light.

Figure 13.13 shows an example of the quality of data that can be obtained from the Kepler spacecraft. It shows a planet (known as Kepler 15b) taking about two hours to pass in front of its parent, a main-sequence star with a temperature roughly similar to the Sun's. As we can read from the vertical axis, the starlight is dimmed by about 1 % during the transit. If we make the reasonable assumption that both the star and the planet have circular cross sections we can conclude that the planet must have a diameter of about 10 % of the star. (See Appendix A.11 for more information about how planetary properties are inferred from Kepler data.) Since we know from its spectrum that the star is roughly similar to our Sun, we can deduce that the planet is about the same size as Jupiter. The planet must be very different from Jupiter, however; whereas Jupiter transits our Sun once every 12 years, Kepler 15b transits its parent star every 4.94 days; making its orbit several times smaller than that of Mercury. Being exposed to so much stellar radiation must give it a temperature far hotter than the Earth and make it quite unlike any planet in our solar system. A number of similar planets have been found, and they have been given the generic name "Hot Jupiters".

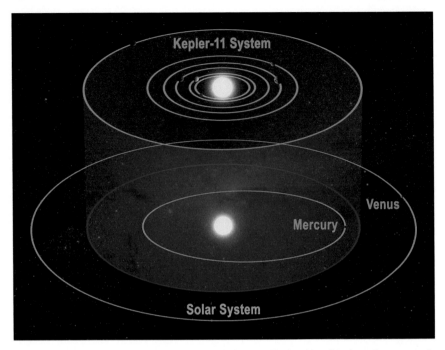

Fig. 13.14 The orbits of the exoplanets around the star Kepler 11, with the solar system for comparison. Image credit: Wikipedia figure by Tim Pyle, NASA

Many early exoplanet discoveries had similarities to Kepler 15b—namely large sizes and small orbits, but this is an example of what scientists call a "selection effect." Within, say, only one year of observations, astronomers cannot possibly find planets with orbits longer than a few months; it is also much more difficult to be sure one has found a transit if the dimming is only 0.01 % than if it is 1 %. Finding small planets in large orbits takes more time and effort than finding large planets in small orbits, but as Kepler discovered more and more transiting planets, a wide range of planetary sizes and orbits emerged.

Kepler has also discovered a number of stars with multiple planets in orbit around them. One of these is Kepler 11 (Figure 13.14) which has at least six planets, all with orbits smaller than that of Venus.

Figure 13.15 shows an analysis of the more than 3,600 candidate planetary systems revealed by Kepler after its first three years of observations. There are several "Hot Jupiters" in the top left hand corner but most of the planets are so-called "Super-Earths" having sizes between those of the Earth and Neptune. Very few Earth-sized planets have yet been found in an orbit as large as the Earth's, but several sightly larger planets have been found with estimated surface temperatures in the range that could support liquid water. It is the search for potentially Earth-like planets that drives most astronomers who work in this area.

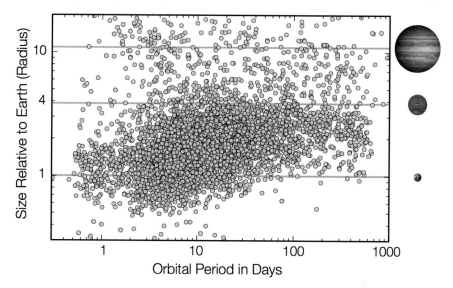

Fig. 13.15 Sizes and orbital sizes of candidate Kepler planets as of 2015. Jupiter, Neptune, and Earth are shown to scale on the right side of the picture. Image credit: Wendy Stenzel and NASA

The Kepler satellite worked perfectly for four years until the 2013 failure of two of the four gyroscopic reaction wheels which Kepler needed to stay accurately pointed in the right direction. It was at first assumed that the mission was effectively over until someone came up with the clever idea of using the Sun's radiation pressure as an aid to keeping the telescope pointed in the right direction. The modified project was given the name "K2" and is still active as of early 2016. In its new configuration the telescope points in one direction for about 80 days then switches to a new direction for the next 80 days. The K2 mission is not quite as sensitive as the original Kepler project, but it has the advantage that astronomers can pick special regions of the sky to study, such as globular clusters and regions of recent star formation. New results are being released regularly by NASA.

NASA is also currently constructing the follow-up mission to Kepler, which is called "TESS" for Transiting Exoplanet Survey Satellite. Its strategy differs from Kepler's in that it looks for transits in front of the 500,000 brightest stars in the whole sky rather than fainter ones in one particular part of the sky. TESS will have four wide-field cameras than can each image a $24° \times 24°$ area of sky. Each part of the sky will be monitored for about a month and any star that shows signs of experiencing a transit will be followed up from the ground. Most of the stars under study are brighter and nearer than those in the Kepler field, so they will be easier to study.

Chapter 14
Accessing Astronomy Surveys

How does one access astronomical surveys? We have moved from clay tablets in the Babylonian era, though handwritten manuscripts, printed books, photographic prints, and CD-ROMS, which, for a while in the 1990s, were distributed to subscribers of professional journals. In recent years, surveys have become so large and the internet has become so accessible that large surveys are no longer disseminated in a complete form. Astronomers use software tools to access the data they are interested in and download it from large servers devoted to the task. Any astronomer, including amateurs, who is willing to learn how to use the necessary software search tools may access these sites and download data from them.

The job of maintaining the computer servers that can provide 24-hour access to these databases has fallen to a small number of institutions that specialize in this service. They are mostly funded by NASA and include:

IPAC The Image Processing and Analysis Center at Caltech in Pasadena, which maintains the data from the IRAS, 2MASS, WISE, Planck, and GALEX surveys. It serves as the repository for the data collected by several other infrared space telescopes including Herschel, Kepler, ISO, and Spitzer. It also houses NED, the NASA Extragalactic Database which merges data on galaxies from multiple observatories, the Keck Observatory archive, and the NASA Exoplanet Archive.

See **http://www.ipac.caltech.edu**

MAST The Mikulski Archive for Space Telescopes based at the Space Telescope Science Institute in Baltimore, which maintains the data from the Sloan and the Pan-STARRS surveys, as well as several other projects such as the Hubble Space Telescope, the International Ultraviolet Explorer (IUE), and the Kepler planet search.

See **https://archive.stsci.edu**

HEASARC The High Energy Astrophysics Science Archive Research Center located in Goddard, Maryland. which maintains data for the COBE, ROSAT, Fermi, EUVE, WMAP, and XMM-Newton missions among others.

See **http://heasarc.gsfc.nasa.gov**

© Springer International Publishing Switzerland 2016
G. Wynn-Williams, *Surveying the Skies*, Astronomers' Universe,
DOI 10.1007/978-3-319-28510-8_14

Galaxy Zoo The Sloan Survey was a pioneer in encouraging amateur scientists to join them in analyzing its data. The human eye and brain is still better than any computer system for noticing subtle patterns in images. A website called Galaxy Zoo was therefore set up to encourage members of the public to help with classifying faint galaxies in the Sloan Digital Sky Survey. Started in 1997 with a dataset made up of a million galaxies it was an immediate immense success. More than 50 million classifications were received by the project during its first year, from more than 150,000 people.

    See **http://www.galaxyzoo.org**

Zooniverse The concept of web-based crowd-sourcing is now being applied in many other scientific studies, led by the Zooniverse website. Anyone with patience and enthusiasm can tag penguins, classify cyclones and compile historical weather records. Astronomy buffs can sit at their desks and explore the surface of the Moon, find planets around other stars, and search for near-Earth asteroids.

    See **https://www.zooniverse.org**

    What all this means is that it is no longer necessary for an amateur astronomer to know the constellations. Or even to own a telescope. All that is needed, rain or shine, is a computer and an internet connection. Happy searching!

Appendices

A.1 Further Reading

Evans, Rhodri, The Cosmic Microwave Background
  Springer, 2015

Hoskin, Michael, The Cambridge Illustrated History of Astronomy
  Cambridge University Press, 1996

King, Henry, The History of the Telescope
  Dover Publications, 1979

Kitchin, Chris, Exoplanets
  Springer, 2011

Longair, Malcolm, The Cosmic Century
  Cambridge University Press, 2006

North, John, Cosmos: an Illustrated History of Astronomy and Cosmology
  University of Chicago Press, 2008

Rowan-Robinson, Michael, Night Vision: Exploring the Infrared Universe
  Cambridge University Press, 2013

Sullivan, Woodruff, Cosmic Noise: A History of Early Radio Astronomy
  Cambridge University Press, 2009

Tucker, Wallace and Giacconi, Riccardo, The X-ray Universe
  Harvard University Press, 1985

Verschuur, Gerrit, The Invisible Universe
  Springer, 2015

Wynn-Williams, Gareth, The Fullness of Space
  Cambridge University Press, 1992

© Springer International Publishing Switzerland 2016
G. Wynn-Williams, *Surveying the Skies*, Astronomers' Universe,
DOI 10.1007/978-3-319-28510-8

A.2 Units

Professional astronomers use the metric system, but very few of us have adopted the strict RMKS standard (meter, kilogram, second); most astronomers still stick to the older cgs (centimeter, gram, second) system when referring to quantities such as a radio wavelength of 21 cm, or an interstellar density of 1 atom per cc.

Other non-RMKS units used in astronomy include "astronomical unit" (AU), which is the average distance between the Earth and the Sun. Another is the "parsec" and its multiples (kiloparsec, megaparsec); the attraction of the parsec as a unit is that in the vicinity of the Sun, stars are typically separated by about 1 parsec.

$$1\,\text{AU} \approx 1.5 \times 10^{11}\,\text{m}$$

$$1\,\text{parsec} \approx 3 \times 10^{16}\,\text{m}$$

$$1\,\text{light year} \approx 10^{16}\,\text{m} \approx 0.3\,\text{parsec}$$

The year (y) is frequently used as a unit of time

$$1\,\text{y} \approx 3 \times 10^{7}\,\text{seconds}$$

Parsecs and years go conveniently together, because of the approximation

$$1\,\text{km/s} \approx 10^{-6}\,\text{parsec/y}$$

The luminosities and masses of stars and galaxies are usually expressed in terms of the luminosity of the Sun ($L_\odot$) and the mass of the Sun ($M_\odot$)

$$L_\odot \approx 4 \times 10^{26}\,\text{watt}$$

$$M_\odot \approx 2 \times 10^{30}\,\text{kg}$$

Special prefixes are often used for large quantities:

kilo   (k) 10^3
mega (M) 10^6
giga   (G) 10^9
tera   (T) 10^{12}

The corresponding prefixes for small quantities are:

milli   (m) 10^{-3}
micro (μ) 10^{-6}
nano   (n) 10^{-9}
pico   (p) 10^{-12}

Infrared astronomers usually define their wavelengths in terms of the micron (μm), which is the same unit as a micrometer.

$$1\,\mu\text{m} = 10^{-6}\,\text{meters}$$

Visible wavelength astronomers, however, generally prefer the Ångstrom (Å), though this unit is not used in this book.

$$1\,\text{Å} = 10^{-10}\,\text{meters}$$

Temperatures in astronomy almost always use the Kelvin scale, sometimes called the absolute temperature. The Kelvin (T_K), Celsius (T_C), and Fahrenheit (T_F) scales are related:

$$T_K = T_C + 273$$
$$T_C = \frac{5}{9}(T_F - 32)$$

A.3 Wavelength and Energy

It is a convention among both physicists and astronomers that X-rays and gamma rays are described by their photon energies rather than by their wavelengths. The energy of a photon (E) is given by the Planck-Einstein relation:

$$E = \frac{hc}{\lambda}$$

where h is Planck's constant, c is the speed of light, and the Greek letter lambda (λ) is the wavelength. Photon energies are almost always expressed in electron-volts (eV), where

$$1\,\text{eV} = 1.602 \times 10^{-19}\,\text{Joule}$$

A photon with energy of 1 eV has a wavelength of 1.24 μm; as the photon energy increases the wavelength decreases.

Radio astronomers sometimes use **frequency** instead of **wavelength** to describe electromagnetic waves, though in this book we stick to wavelength for consistency. Frequency, which is usually expressed using the Greek letter nu (v), and wavelength lambda (λ) are related to each other by the equation:

$$v = \frac{c}{\lambda}$$

where c is the velocity of light. A frequency of 5 GHz corresponds to a wavelength of 6 cm; as the frequency increases the wavelength decreases.

A.4 Diffraction Limit

The resolving power of a telescope is a measure of its ability to see fine detail—specifically its ability to separate the images of two celestial objects that are separated by only a small angle on the sky. As discussed in section 1.2, there are a number of factors that can affect the sharpness of an astronomical image, but the ultimate limit, named the diffraction limit, is set by the laws of optics; to see fine detail in an image it is necessary to maximize the ratio of the diameter of the primary mirror (D) to the wavelength (λ) of the radiation being measured.

If there are no other causes of blurring, a point source on the sky will appear to have a beamwidth of approximately θ_d, where

$$\theta_d = 4200(\lambda/D)$$

where D and λ are measured in the same units, and θ_d is measured in minutes of arc ($1/60$ of a degree).

Table A.1 shows some examples of diffraction limits for sample telescopes at visible, infrared, and radio wavelengths. The final column lists whether the diffraction limit is usually achieved with these instruments in practice. Generally speaking observations with radio telescopes, especially those which make use of aperture synthesis, are diffraction limited, while those at shorter wavelengths are not. For ground-based optical telescopes the resolving power is often set by the blurring caused by the Earth's atmosphere; for X-ray telescopes it is usually set by aberrations in the optics.

Telescope	Diameter	Wavelength	Diffraction Limit	Achievable?
Human Eye	5 mm	0.6 μm	0.5 arcmin	Yes
Amateur Telescope	25 cm	0.6 μm	0.6 arcsec	No
Hubble Space Telescope	2.4 m	0.6 μm	0.06 arcsec	Yes
Palomar Observatory	5.0 m	0.6 μm	0.03 arcsec	No
Planck Satellite	1.5 m	3 mm	8 arcmin	Yes
Parkes Radio Telescope	64 m	75 cm	50 arcmin	Yes
Very Large Array	36 km	21 cm	1.5 arcsec	Yes

Table A.1 Diffraction limit examples

A.5 Precession

Precession of the equinoxes occurs because the Earth is not a perfect sphere; centrifugal forces arising from the Earth's daily rotation make it slightly fatter across the equator than between the poles. It can be shown by an application of Newtonian physics that the interaction between the Earth's oval shape and the gravitational field of the Sun causes the direction of the Earth's rotation axis to slowly change (Figure A.1). Right now, the Earth's rotation axis points close to the star Polaris, but over a period of 26,000 years it will move in a circular path that is 23.5° in diameter centered on a point in the constellation of Draco. About 12,000 years in the future the Earth's rotation axis will point roughly in the direction of the star Vega.

The existence of precession means that catalogs of star positions must always specify the epoch for which that position is valid, Standard epoch dates include January 1 of 1950 and of 2000. Astronomers must calculate a correction to a star's published position if they want to observe it on any other date.

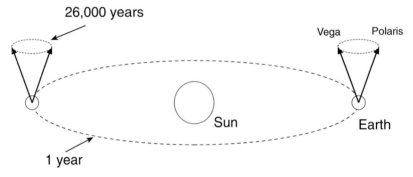

Fig. A.1 Precession of the Equinoxes

A.6 Stellar Magnitudes

It was Hipparchus who introduced the idea of stellar magnitudes—an idea that is familiar to all astronomers. He labeled the brightest stars as 1st magnitude and the faintest ones as 6th magnitude. Since Hipparchus had no way of measuring the amount of light that a star emitted, his system was subjective.

Following the development of stellar photometry in the 19th century, however, the British astronomer Norman Pogson proposed setting up a logarithmic system in which a 5 magnitude brightness difference corresponds to a factor of exactly 100 in received power. In this scheme, which is still in use today, one magnitude corresponds to a brightness ratio of $100^{1/5}$ or 2.512. For visible light, the star Vega is defined as zero magnitude.

As an example, we have noted in section 12.2 that the faintest stars in the Pan-STARRS photometry catalog are 22nd magnitude. The power we receive from such a star is $(2.512)^{22}$ or nearly a billion times less than we receive from Vega.

A.7 Coordinate Systems

The three most widely used astronomical coordinate systems are the equatorial coordinate system, which is tied to the rotation of the Earth; the ecliptic coordinate system, which is tied to the Earth's motion around the Sun; and the galactic coordinate system which is linked to the Milky Way.

Ecliptic coordinates are the natural system to use when one's main interest is in following the motion of the Sun, Moon, and planets against the background of stars; hence their use in early catalogs. They also have the advantage that corrections for precession are easy to calculate and apply.

As stellar navigation became more important and more precise in the 17th century, catalogs like that of John Flamsteed (section 3.2) started to use the Earth-based equatorial coordinate system instead of the ecliptic coordinate system. This made navigational calculations much easier to perform.

A second advantage of the equatorial coordinate system is related to the design of telescope mounts. The nineteenth century developments of photography and spectroscopy, with their need to keep a telescope accurately trained on one patch of the sky for minutes or hours, encouraged the adoption of equatorial mounts for many telescopes. The UKIRT telescope, shown in Figure 6.10, is an example of an equatorially-mounted telescope. With this kind of mount, a star can be accurately tracked by a steady clock-driven increase of the hour-angle, with no change needed in the second axis—the declination. Finding and tracking a particular star with an equatorially-mounted telescope is easy if the star's right ascension and declination are known.

The third system, galactic coordinates, was introduced at a later date and became increasingly common during the 20th century. The main reason for its popularity is that surveys at non-visible wavelengths, such as radio, infrared, and X-rays, typically see to much greater distances from the Sun than do surveys at visible wavelengths, which are dominated by nearby stars. Surveys in these wavelength bands often contain a rich mixture of galactic and extragalactic objects, and maps in galactic coordinates make it easier to distinguish between these different kinds of sources.

All the full-sky maps in Chapters 5 though 12, except for Figures 5.4 and 6.4, are plotted in galactic coordinates.

A.8 Interstellar Extinction and Reddening

As we discussed in Chapters 5 and 6, the space between the stars in the Milky Way Galaxy is filled with interstellar matter, some of which is concentrated into clouds or nebulae, and some of which is spread out more or less uniformly. The most common ingredients of this interstellar medium are hydrogen and helium gas, but about 1 % by mass comes in the form of microscopic dust particles, including graphite, ice grains and silicates. It is these dust grains that absorb and scatter light, causing distant stars to appear dimmer to us than they would if interstellar space was empty.

The density of interstellar matter is highly variable, but in the plane of the Milky Way starlight is attenuated by something like one magnitude (a factor 2.512) for every 500 parsecs the light travels through it. This extinction is compounded as the light travels farther, so that after 5,000 parsecs the light is dimmed by a factor 2.512^{10}, or 10,000. This is why it is so difficult to study anything other than the local few hundred parsecs of our Galaxy using visible light.

Because dust grains are so small they behave differently towards electromagnetic radiation of different wavelength; long wavelength light (red) is less affected by dust than short wavelength light (blue) which leads to what is called interstellar reddening: distant stars appear to have a redder color than similar nearby stars. By the same token, infrared light is less affected by interstellar dust than visible light: specifically, 2.2 µm infrared radiation suffers only 10 % as much extinction as visible light with a wavelength of 0.5 µm. The 2.2 µm radiation from the 5,000 parsec distant star in the previous paragraph therefore suffers only one magnitude of extinction, not ten magnitudes.

A.9 Radiation, Temperature, and Wavelength

One of the first breakthroughs of quantum physics was the derivation of what we now call Planck's Law, which describes the nature of the electromagnetic radiation emitted by a hot object. Implicit in this theory is the idea of a Black Body, in the sense of a surface which completely absorbs 100 % of the electromagnetic radiation incident on it. The relationship that Max Planck derived in 1901 is shown in Figure A.2 for three different temperatures. As the temperature increases, the emission gets stronger at all wavelengths, with the peak of the emission moving to shorter wavelengths. A body which is not perfectly "black" will emit less radiation than is indicated by the following graph; the extent of the deficit need not be the same at all wavelengths.

The relationship between the peak wavelength and the absolute temperature was already known experimentally before Planck's time, and carries the name "Wien's Law". It states that the wavelength at which the object emits the most radiation (λ_m) depends on its absolute temperature (T) according to the equation:

$$\lambda_m T = 3000$$

where λ_m has units of microns, and T has units of Kelvins

Wien's law provides an explanation for the differences in the colors of stars: the white-colored Sun, which has a surface temperature of about 6000 K, emits most strongly at a wavelength of 0.5 μm, in the middle of the visible spectrum. The much redder star α Centauri, with a surface temperature of 4400 K, emits most prominently at 0.7 μm, right at the long-wavelength (red) end of the visible spectrum. Wien's Law is applicable to much more than just light. The table below shows some of the phenomena which are discussed in this book, with their associated approximate values of λ_m and T.

Object	T	λ_m
Cosmic Background radiation	3 K	1 mm
Molecular Clouds	100 K	30 μm
Earth	300 K	10 μm
Sun	6000 K	0.5 μm
X-ray accretion disk	1,000,000 K	3 nm

Table A.2 Astronomical examples of Wien's Law

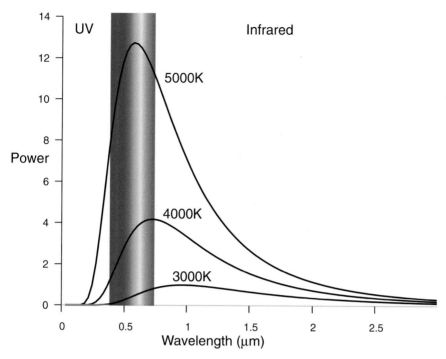

Fig. A.2 Radiation from a Black Body at three different temperatures. Adapted from Wikipedia diagram by Darth Kule

A.10 Redshift-Age Relationship

In physics, the Doppler shift of a spectral line is the apparent change of its wave-length that is caused by the relative motion of the source and observer. When they are moving away from each other the wavelength appears lengthened and is said to be redshifted; when they are moving towards each other the wavelength appears shortened and is said to be blueshifted.

The redshift (z) is defined by the following equation

$$z = \frac{\lambda_{obs} - \lambda_0}{\lambda_0}$$

where λ_{obs} is the observed wavelength of the spectral line, and λ_0 is the intrinsic, or rest wavelength of the spectral line.

For separation velocities (V) that are small compared with the velocity of light (i.e. $z \ll 1$), we can write

$$z \approx \frac{V}{c}$$

In 1929 Edwin Hubble discovered that many more galaxies had redshifts than blueshifts, and that the redshift of a galaxy is proportional to its distance from us. This discovery, called Hubble's Law, lent strong support to the now generally accepted idea that the universe as a whole is expanding as it gets older. It also provided astronomers with an extremely powerful method of determining the distance (D) to a galaxy from the measured recession velocity V, using the formula

$$V = H_0 D$$

where H_0 is called Hubble's constant. The current best value for H_0, determined by the Planck mission (section 7.4), is 67.8±0.9 km/sec per megaparsec.

The light from distant galaxies can take a significant amount of time to reach us, so observing galaxies with large redshifts is equivalent to viewing the universe as it was when it was much younger than its current age of 13.8 billion years. If we make some well accepted assumptions about the density and geometry of the universe, we can use Einstein's general theory of relativity to calculate the relationship between redshift and age. This relationship is shown in Figure A.3. Galaxies with redshifts of around 1, for example, emitted the light by which we see them when the universe was only 5 billion years old, less than half of its present age. Galaxies with redshifts of greater than 6 are seen as they were when the universe was less than 1 billion years old, or 7 % of its present age.

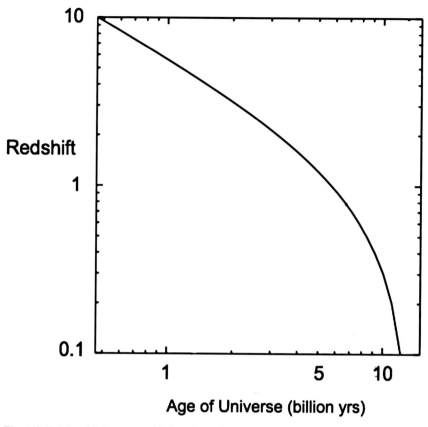

Fig. A.3 Relationship between redshift and age since the Big Bang

A.11 Determining the Properties of an Exoplanet

A remarkable amount of information can be obtained about an exoplanet by combining data from a transit experiment like Kepler (section 13.4) and Doppler radial velocity measurements from a ground-based telescope. All that is needed to interpret the data is a modest amount of basic Newtonian physics.

Let us assume that a planet has been observed by Kepler to be regularly transiting its parent star and, for simplicity, that ground-based observations of the radial velocity of the star show that the planet's orbit is circular. The existence of a transit tells us that the plane of the orbit goes through our line of sight rather than being tilted at some unknown angle.

Let $M_\star$ and M_p be the masses of the planet and star, and $R_\star$ and R_p be their orbital radii around their common center of mass. Balancing the gravitational attraction between them with the centrifugal force holding them apart gives us

$$\frac{GM_pM_\star}{(R_p+R_\star)^2} = M_p\omega^2 R_p$$

where ω is the angular velocity of the planet around the star. Since it is reasonable to assume that the mass of the star $M_\star$ is much larger than the mass of the planet M_p we can make the approximation that $R_p \gg R_\star$. Noting also that $\omega = 2\pi/T$, where T is the orbital period of the planet, we get

$$GM_\star = \left(\frac{2\pi}{T}\right)^2 R_p^{\ 3} \tag{A.1}$$

If we know the spectral type of the star we can make a good guess at its mass $(M_\star)$ and thus derive the radius R_p of the hidden planet's orbit from the measured orbital period of the star by inverting equation A.1:

$$R_p = \sqrt[3]{\frac{GM_\star T^2}{4\pi^2}} \tag{A.2}$$

Now consider the conservation of momentum as the star and planet orbit their common center of mass with velocities $V_\star$ and V_p

$$M_\star V_\star = M_p V_p = M_p \left(\frac{2\pi R_p}{T}\right)$$

so that

$$M_p = M_\star \left(\frac{V_\star T}{2\pi R_p}\right)$$

Since we know T from measurements of the separation of successive transit events, $V_\star$ from the maximum Doppler velocity as the planet orbits the star, and have derived R_p from equation A.2, we can calculate the mass of the hidden planet.

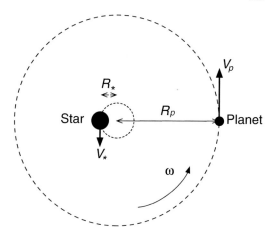

Fig. A.4 A simple
star/exoplanet system

This is possible even if the planet is invisible and even if we do not know the distance to the parent star.

We can also calculate the radius (r_p) of the transiting planet by using the Kepler data to measure by how much the light of the star is dimmed during a transit. For example during the transit shown in Figure 13.13 the planet hides 1 % of the surface of its parent star. Since the cross sectional areas of the star and planet are proportional to the squares of their radii we can deduce that the radius of the planet is $\sqrt{1\%}$ or 10 % the radius of its parent star. This is how it was deduced that Kepler 15b was about the size of Jupiter.

Since we now know both the mass of the planet and its radius we can calculate its density ρ from the equation

$$\rho = \frac{\text{Mass}}{\text{Volume}} = \frac{3M_p}{4\pi r_p^{\,3}}$$

In this way we can get an idea of whether the planet is rocky, with a high density, or gaseous, with a low density.

Although the light from any star will greatly outshine that from any of its associated planets, it should be remembered that what we measure during most of an orbit is the *sum* of the light from the star and the planet. In some cases Kepler has been able to measure the drop in brightness when the planet is eclipsed by moving *behind* its parent star. If we can measure the size of this effect we can calculate the surface brightness of the planet since we know its radius. Its surface brightness at visual wavelengths can give us an idea of the reflectivity of the planet's surface, while its surface brightness in the infrared waveband can give us an estimate of its temperature.

Note that all of these parameters, except the last one, the surface brightness, have been determined without detecting any radiation from the planet itself.

Index

Symbols

21-cm line, 47, 59, 119
2MASS survey, 3, 73, 83, 165
3C survey, 55
8C survey, 57

A

ACBAR experiment, 91
Accretion disk, 110, 111
Adaptive optics, 4
AFGL survey, 76
al-Sufi, Abd al-Rahman, 14
Albedo, 82
ALEXIS survey, 103
Allen Telescope Array, 67
Almagest, 11
Alpher, Ralph, 87
Andromeda galaxy, 14, 140
Angular resolution, 4, 171
Apache Point Observatory, 136
Aperture synthesis, 55, 91, 171
ARAKI satellite, 81
Arecibo Observatory, 66, 67
Argelander, Freidrich, 35, 133
Aristotle, 10
Armillary sphere, 11
Asteroids, 24, 82, 140, 142
Astrographic Catalogue, 38, 131
Astrometry, 129
Astronomical unit, 168
Astrophotography, 37
ATLAS survey, 144
Atmospheric seeing, 4, 129
Atmospheric transmission, 47
Aumann, George, 80

B

Babylonian astronomy, 9
BATSE experiment, 122
BD catalog, 35
Beamwidth, 4, 171
Bell, Jocelyn, 57
BEPPOSAX satellite, 5, 123
Bessel, Friedrich, 35
Beyer, Johann, 19
Big Bang, 54, 87
Binary stars, 34, 107, 110, 129
Black body, 87, 89, 176
Black hole, 110, 111, 123, 152, 156
Blazar, 120, 124
Bode, Johann, 24
Bolometer, 92
Bonn radio telescope, 50
Boomerang experiment, 91
Brahe, Tycho, 22, 32

C

Cannon, Annie Jump, 46
Carte du Ciel, 38, 131
Cassegrain telescope, 112
Cassini, Giovanni, 24
Cassiopeia A, 49, 53
Catalina Sky Survey, 143
CCDs, 135
Center of mass, 180
Cepheid variable, 32, 131, 139
Chajnantor Observatory, 91
Chandra Observatory, 115
Chelybinsk impact, 82
Cherenkov radiation, 124
Chinese astronomy, 16

© Springer International Publishing Switzerland 2016
G. Wynn-Williams, *Surveying the Skies*, Astronomers' Universe,
DOI 10.1007/978-3-319-28510-8

Printed in the United States
By Bookmasters